Witness To A Changing Earth

C. Hans Nelson

Witness To
A Changing Earth

A Geologist's Journey Learning About
Natural and Human-caused Global
Change

 Springer

C. Hans Nelson
Granada, Spain

ISBN 978-3-030-71813-8   ISBN 978-3-030-71811-4   (eBook)
https://doi.org/10.1007/978-3-030-71811-4

This Springer imprint is published by the registered company Springer Nature Switzerland AG
The registered company address is: Gewerbestrasse 11, 6330 Cham, Switzerland

*To my family, friends, colleagues and readers in the hope
you can work towards mitigating the effects of global change.*

Foreword

Earth will survive our human-caused limiting factors, but will the human species?

Hans Nelson is well known in the global marine geology community for his pioneering studies on turbidite systems off continental margins as records of past earthquakes, along with potential hazards and petroleum reservoirs. He is the quintessential geologist, seeking to understand how our Earth System functions by studying records of its past, mainly beneath the sea floor using corers and seismic gear but also at times in lakes, and in outcrops on land for the more distant past.

Throughout a long and distinguished career he has worked with many colleagues on projects large and small from the Arctic to the tropics, with his findings well published and widely read. He has also noticed, like many other geoscientists, the many signs of change in the Earth System—atmosphere, biosphere, ice, land, and oceans, as Earth's human population rises and the demand for resources increases. We know this is unsustainable. Demand must be reduced at every level in human society if increasingly severe catastrophes are to be averted.

In this book, Hans draws on his lifetime of study and travel to share with others his experience of the natural world, because he, like many other scientists, has become deeply concerned at the growing degradation of our environment. Part I begins in the first section with close encounters in Minnesota with lightning and floods. He then introduces the reader with his own experience of first the natural and unavoidable hazards of floods, earthquakes, and volcanoes, before going on to consider the human-induced hazards we are bringing about ourselves in global warming and ocean acidification, and the terrible consequences of both natural and human-induced combined.

In Part II, Hans expands to educate the reader about climate and global warming from many lines of evidence linking increasing CO_2 and methane in the atmosphere to rising ocean temperatures and sea level, as well as shrinking polar and mountain ice in addition to permafrost. He also reviews increasing extinction rates and separates natural climate variability from human-caused global warming, and recognizes contamination and pollution problems unrelated to greenhouse gases.

Part III of this book sets this book apart from most others with Hans' extensive review based on his personal experience of solutions for these growing global change problems, and his advocacy for a shift in attitude through education at all levels focused on a sustainable way of life for humanity and Earth itself.

Hans concludes "The bottom line is that our future outlook can improve with a change in attitude for the common good of humanity and health of Earth, particularly attitudes of populist politicians, the wealthy few and powerful multi-national corporations. Our choice must be to put the common good ahead of greed and short-term economic goals." His insights from a lifetime of scientific investigations covering all aspects of Earth System workings, along with his observations on the failings of the West in particular, and his many stories of experiences with fellow humans provide readers with compelling reasons and inspiration for attitudinal change.

Peter Barrett, RSNZ, NZAM, FGS (Hon)
Fellow of the Royal Society of New Zealand
Holder of the NZ Antarctic Medal
Honorary Fellow, Geological Society of London
Emeritus Professor of Geology
Victoria University of Wellington
Wellington
New Zealand

Preface

.

The most important reason for me to write this book and my greatest concern is the multiple types of human-caused global change that are accelerating rapidly, threatening the quality of life and even the existence of humanity on Earth. Many of the media and government reports equate human-caused climate change or global warming with global change. Unfortunately, humans cause many other types of global change that are occurring at the same time as warming. Most people are aware that humans are introducing carbon dioxide (CO_2) and methane (CH_4) into the atmosphere and these gases are warming the global climate and ocean. However, in addition to warming the ocean, which melts polar ice sheets and destroys coral reefs, humans are trashing the ocean with plastic that in a few decades will weigh more than the fish in the ocean. Similarly, on land humans are warming, eliminating, and polluting habitats, which results in animal extinctions equal to natural extinctions that occurred earlier in geological times, when humans were not present. We humans need to remember that we too are animals facing possible extinction as a result of our global changes.

In this book, I use the term global change to describe worldwide natural as well as human-caused events, because both types can have global effects on our Earth. My descriptions of natural global change focus on earthquakes and volcanoes, because these two natural types of geologic events result in the most widespread global change and I have spent much of my career investigating these events. Great earthquakes of magnitude (Mw) 9, like the one in Sumatra in 2004, result in tsunamis that spread globally across entire ocean basins, and cause hundreds of thousands of deaths for distances of more than 1000 km (600 miles) along coasts and inland where the earthquakes occur. Catastrophic volcanic events like Crater Lake Oregon, which I have spent a lifetime studying, spread ash clouds that cover the Earth, cool the climate for years, and also cause thousands of deaths. My interest in investigating natural global change is to help mitigate the risk posed by these natural events.

These aforementioned few examples of natural and human-caused global changes are some of the many that I describe throughout the book. Because human-caused rather than natural geologic global changes are now dominating the Earth's environment, the present geologic time is called the Anthropocene (https://en.wikipedia.org/wiki/Anthropocene). Humans have become a force of Nature changing the Earth System, just as volcanism and glacial cycles do. For example, when I studied the Ebro River drainage system, the largest on the Iberian Peninsula, I found that humans caused the same effect as the changes of the Pleistocene glacial climate (from 2.6 million to 11,700 years) and Holocene warm climate (the last 11.700 years). Likewise, a biological study of mine has shown how the Pacific walrus population in the Arctic has been reduced by 50% because of global warming and loss of their sea ice habitat. These and other scientific discoveries on my geologist's journey have provoked me to alert the public about the multiple human-caused global changes in addition to climate change.

The force that drives all of the human-caused global change is population growth. The Earth's human population has tripled just during my lifetime. This exponential increase has led to the many global changes that I will describe. The huge increase in the population just in the last few decades also has resulted in a dangerous interaction between both natural and human-caused global changes. For example, large numbers of people have moved into coastal areas that are subject to great earthquakes and tsunamis such as the Pacific Northwest, California, and Japan. People also have crowded into areas around the world that are subject to hurricanes, typhoons, and flooding by rising sea level and higher storm surges. The 2020 typhoon in Bangladesh, which crowded refugees together during the COVID-19 pandemic, shows how natural and human global change can be interconnected. Unfortunately, the development of infrastructure has not kept up to protect the rapidly growing population from both natural and human-caused global changes.

Humans cannot prevent the natural events like earthquakes, but we can survive them, whereas we have the ability to alter human-caused global changes and keep a sustainable planet. For example, I have experienced an unusual number of dangerous natural events, but also have been lucky enough to survive them. While growing up I experienced a tornado, a flood, and several lightning strikes. Later in my life I have experienced several earth-quakes and catastrophic storms, which had winds up to 300 km per hour (180 mph) and waves up to 25 m (80 feet) high. My worst earthquake experience was the 1989 Loma Prieta earthquake in the San Francisco area. I thought we would lose our house because downed power lines had set a fire nearby and the main water line to the area had broken. Fortunately, the fireman arrived in time to put out the fire, and because of my previous encounters with earthquakes, I had designed my California house for earthquakes. However, others were not so lucky and several houses within a kilometer of mine were destroyed. My house example shows that we can plan for and survive natural catastrophes. Also, you can see how all these experiences with natural-type global change have provoked thoughts for my book.

As well as my experiences with natural global change, I have had many personal experiences with human-caused global change that also have provided background for my book. I have observed pollution in Lake Baikal Russia from the world's largest paper mill and seen recent coral bleaching while skin diving on Malaysian coral reefs. Now when swimming off the Costa Brava in northern Spain, because of the warming and acidified Mediter-ranean Sea, masses of jellyfish often sting you compared to swimming in the 1980s. Also while living in Spain during the past 20 years, I have seen African dust storms from expanding deserts become more severe. The fruit in our garden ripens and flowers bloom a month earlier. Our semi-tropical jacaranda tree, which never bloomed the first decade living in Spain, now blooms every year.

In addition to my widespread global experiences, how has my scientific background qualified me to write about global change? My first B.A. degree from Carleton College was with a major in geology and a minor in biology. Likewise, my M.S. degree from University of Minnesota was in geology and biology where I studied effects of glacial climate change and the geological limnology of Crater Lake, Oregon. My Ph.D. was in geological oceanog-raphy from Oregon State University where I also took every marine biology course taught including phytoplankton (microscopic plants), zooplankton and foraminiferal ecology (microscopic animals), nekton (marine fish and mammals), and benthic biology (the study of life on the sea floor). In sum, I consider myself to be an Earth scientist or marine geologist/biologist and have been trained in geology, biology, and oceanography.

From the title you may think that as a geologist I just look at rocks. However, as you will see from examples of my studies, I have examined the shape of the land, lake floors, and ocean seafloor surfaces; the plants and animals in the water and sediment at the bottom of lakes and oceans; the history of this sediment; and the history of ice on land and in the sea. The reason that geological oceanographers or marine geologists like me are broadly trained environmental scientists is that we apply all the sciences to the study of lakes and oceans. We look at the physical oceanography of ocean circulation and currents; the chemistry of ocean and lake water; the biology of the water and sediments; and the geology of coasts, seafloor, and sediments.

My first global experience was teaching marine biology on the University of the Seven Seas and helping lead field trips in all of the temperate and tropical environments of the northern hemisphere. My professional onland and aquatic studies have taken place in the Arctic, Antarctic, Atlantic, and Pacific Oceans, Mediterranean, Bering, and Chukchi Seas, as well as Minnesota Lakes, the volcanic Crater Lake in Oregon, and the Rift Lake Baikal in Russia. This research has focused on resources, geologic hazards, and environmental assessment. My studies have included work for government agencies, universities, and industry. This breadth of academic training and research experience gives me a broad perspective on different viewpoints about global change.

For years, I have wanted to alert people to the challenges of global change and write a book based on my worldwide experiences from the last 60 years. So, why should you want to read this book? In this book, you will learn about how geologists study natural global changes such as earthquakes and volcanoes, and the adventures that happen during these investigations. You also will learn about human-caused changes of global warming; resource depletion; and pollution of air, water, and soil. Readers will find out that if we continue the present human habits for the next 50 years, the Earth faces dramatic decrease in water quality and quantity, farmland, wild marine fisheries, forests, fossil fuel supplies, and plant and animal species.

The general public, teachers, and students may wish you knew what to do about these human-caused changes, because our Earth's population continues to increase along with demand for those ever-decreasing resources. To counter these fears, my book describes solutions and examples of successful programs combating global change. By the end of this book, readers and policy-makers will understand how countries can learn to avoid and prevent catastrophic loss of life and property associated with natural and human-caused global changes. People and policy-makers also will find examples of global technical and political advances to achieve the goal of a sustainable Earth.

As well as the scientific purpose, I hope that this book will spark your interest in the natural world by showing how exciting studying nature can be. I hope readers will see that natural science studies can be as adventurous as any action movie or as humorous as a comedy. We need to encourage youngsters to spend less time in the virtual world of computers and more time enjoying the real natural world. Maybe even some young readers will be inspired to become natural scientists.

To my knowledge, this is the only book you will find that describes both natural and human-caused global changes while showing the gravity of these combined changes occurring at the same time. Although much of this book describes the dangers of the combined effects of global change, at the end I point out solutions for global change problems and a sustainable Earth. My desire is that the book will empower you to make positive changes for global change problems and the severe challenges humans face to keep a sustainable planet.

I hope my book will inspire the readers to adapt the mantra:

MAKE EARTH GREAT AGAIN!

Granada, Spain   C. Hans Nelson

Acknowledgements

There are many people to thank for helping me on my geologist's journey. My family has sacrificed the most because of my long absences for oceanographic cruises and scientific meetings. Some years my wife Carlota Escutia and I have only been at home together for as little as 2 months, because of our combined scientific activities. My older daughters Laurie and Lisa endured a divorce that resulted from months at sea for many years in the Arctic. My younger daughters Carli and Cristina also saw little of me in their earlier years. I cannot remember which of my young daughters once said that if dad dies it will just seem like he has gone on another scientific cruise.

I never would have embarked on my geologist's journey without the world-class mentors and teachers that guided me through my education and into my career. My high school teacher Bob Hanlon in the early 1950s was way ahead of his time in teaching ecology with his advanced biology course. My undergraduate geology advisor, Dr. Eiler Henrickson, not only led me to change my major to geology, but also taught me lifelong lessons for sports and physical exercise. Dr. Herb Wright, my MS advisor at University of Minnesota, encouraged my lake studies through his pioneering limnology group. My Ph.D. advisor, Dr. John Byrne, was the best teacher that I ever encountered, and allowed me to focus on my thesis research rather than his interests. Dr. David Hopkins, my first project chief at USGS, taught me about Alaskan geology and how to achieve scientific results in the logistically difficult remote areas of Arctic Alaska.

There are many scientific colleagues and friends who are too numerous to mention for their help with my scientific research and my family during my absences for oceanographic cruises and onland fieldwork. None of my scientific research could have been achieved without the help of hundreds of other scientists and technicians in laboratories on land and ships at sea. For example, I have had as many as 40 scientists and technicians on a scientific cruise and 35 scientists in my USGS/CSIC project in Spain. Also my marine geology projects could not have been completed without hundreds of shipboard officials and crew from USGS, NOAA, ONR, USNPS, Oregon State University, CSIC Spain, IFREMER France, Texas A & M University, Scripps Institute of Oceanography, and Academy Nauk, Russia.

Among many, Dr. Peter Barrett, Patric Coupland, Ranny Eckstrom, Dr. Carlota Escutia, Dr. David Lowe, Duncan MacMillan, Carli Nelson, Wim Velsink, and the editor Dr. Alexis Vizcaino have been, especially, helpful with my book efforts by providing scientific advice and reviews.

Contents

About the Author

C. Hans Nelson is a guest scientist at Instituto Andaluz de Ciencias de la Tierra in Granada, Spain, and a Research Associate for the Turbidite Research Group at University of Leeds, Great Britain. During his career in government, academia, and industry, he has carried out research on marine geologic hazards, resources, and environments mainly at the U.S. Geological Survey as well as at research institutes and universities in USA, Spain, France, and the Netherlands. This has resulted in more than 200 peer-reviewed articles and books, and the Geological Society of America Kirk Bryan Award for Research Excellence. He has taught short courses on deep-sea deposits worldwide, for governments, geologic societies, universities, and industry. Hans has served as a Chief Scientist for 30 scientific expeditions, in Bering, Chukchi, Pacific,

Atlantic, Gulf of Mexico, and Mediterranean Seas as well as Crater Lake, Oregon, and Lake Baikal, Russia. Photo courtesy of the author, taken while a lecturer on a cruise ship in Glacier Bay, Alaska, July 2000.

1

Introduction

I write this book because along my journey as a land and sea-going earth scientist I have witnessed natural geologic and human-caused global changes that have, or might have, a profound effect in human lives. Natural disasters can cause global changes that are as significant as human-caused changes, including climate change. Wikipedia defines a natural disaster as a sudden event that causes widespread destruction plus significant loss of life and collateral damage caused by forces other than the acts of human beings. Natural disasters include events such as earthquakes, floods, volcanic eruptions, landslides, and hurricanes. These disasters can cause profound global change effects and I will focus on volcanic eruption and earthquake changes that I have studied. However, the main focus of the book will be on human-caused global change.

Look around and you see signs everywhere of what humans have done to our Earth. Most of you have also seen news reports, prior to the coronavirus pandemic, of countries where people are wearing breathing filters and the sky is not visible because of smog. You have heard reports about how we pollute our drinking water supplies. You may have felt sadness every time you hear about another of our large wild animal species going extinct. We are losing productive farmland. Summers are hotter and winters colder than when we were young.

Many of these natural (e.g. Sumatra 2004 and Japan 2011 earthquakes) and human-caused global changes have occurred just during my lifetime as an earth scientist. The human-caused global changes give me grave concern

C. H. Nelson, *Witness To A Changing Earth*, https://doi.org/10.1007/978-3-030-71811-4_1

1

because I have children and grandchildren that will have to face the consequences of these changes and live in a diminished world lacking the beauty and joy of experiencing the natural environments. My family also will have to live with the natural geologic hazards in their areas and have better protections against them.

Science can provide some answers to these problems, but citizens of the world have to have the political will as well, as has been starkly shown by the 2020 Covid-19 pandemic. For along with any improvements for global change problems is a cost; these may affect global economics and individual living standards. There needs to be a broadened world-view to include not only our own country, but all others as well. These issues are outside the realm of science to address.

Many people try to consider both the scientific and non-scientific aspects of human-caused global change and get caught up on one side or the other causing a large chasm between each other when trying to find solutions. These points of view need to come together for the greater good of our entire Earth.

How does the average person distinguish between science and non-science and recognize the roles of each? Science works by collecting data, the more the better. This often is limited by money and time available to collect the data and we need to be sure there is sufficient funding to collect critical data to understand human-caused and natural changes to our Earth. Scientists review each other's data and find flaws with it. If resources are available, other scientists may try to duplicate and thus verify or refute the data by collecting their own data. Scientists may then develop a "theory", to explain the data, and again this may undergo repeated review and criticism. The theory may then be used to predict what would be expected in similar circumstances.

In contrast, many non-scientific arguments lack data and continued review and scrutiny. It often is emotion-based opinion based on effects to the individual, such as loss of income, increase in production costs, loss of individual freedom etc. So to distinguish between whether a person, or point of view is scientific or not, ask yourself if there has been a rigorous collection of data, scrutiny by many individuals knowledgeable in the field, review and correction, and agreement by the majority of scientific specialists that the data and theory are correct. In this case, science can provide the direction for a solution. If a point of view is not scientific, the arguments will lack background data, a reliable cause and effect relationship in a theory, and responses to questions are repeated mantras. In this case, the response to global change problems requires a more emotional response to recognize and address the non-scientific concerns. As Nate Silver has noted, human-caused climate skeptics cannot just rummage through fact and theory alike for ideological convenience, but must weigh the strength of the new evidence against the overall strength of the theory (Silver [3]). Cherry-picking scientific data to suit your beliefs is not an option to solve global change problems.

A great deal of the loss of human life and property can be avoided if global scientific studies are conducted and applied to warn people and plan for natural catastrophes. Similarly, human-caused global changes do not have to take place if humans unite, take intelligent control of our lives and choices to evolve towards a sustainable use of the earth's resources for future generations. This evolution will be most important for the developed countries that use the majority of the resources. Underdeveloped countries already face these oncoming changes. There are conflicts related to resources, the worst catastrophes take place related to extreme weather conditions of severe hurricanes, floods, and droughts, and these countries suffer from a lack of scientific knowledge about natural catastrophes.

An important purpose of this book is to point out that because many humans now live in cities in the developed world, there is an increasing disconnect with the natural world compared to a century ago when most people lived on farms. The rapidly advancing technology has resulted in people believing that technology can solve all of the global change problems. However, we still live in the natural world and need air to breath, water to drink and soil to grow our food. We still face natural catastrophes such as the recent earthquakes and tsunamis in Sumatra (2004) and Japan (2011) or floods in Houston (2017), hurricanes in the Caribbean (2017), and wildfires in western North America (2017, 2018, 2020). Some of these catastrophes are bringing human development into conflict with known natural processes. For example, the destruction of the Fukushima nuclear power plant occurred in the 2011 Japanese earthquake, even though earth scientists had warned that the world's strongest type of earthquakes were possible and they had shown that similar tsunamis struck these power plant sites about every 800 years (e.g. (Minoura et al. [2]; Goldfinger et al. [1]).

An additional purpose of the book is to show the importance of understanding both natural (e.g. 2011 Japanese earthquake) global changes in Part l and human-caused changes (e.g. global warming) in Part ll. The Part l memoirs and Part ll also will show examples of how earth scientists do their studies of natural and human-caused changes that are so important for human safety and a sustainable earth. In Part lll solutions for global change problems and a sustainable earth are suggested.

A bottom line purpose for the book, as shown in Part III, is that we need public education and political interest to support natural science investigations to assess natural and human-caused global changes so that we can reduce unnecessary deaths and find sustainable ways for the human population to continue to survive on our Earth. At present, the budget cuts since the 1970s have reduced USA federal research in science. For the most recent example, note how the budget cuts have hurt the health sciences and compromised the USA ability to cope with the covid-19 pandemic.

References

1. Goldfinger C, Ikeda Y, Yeats RS, Ren J (2013) Superquakes and supercycles. Seismol Res Lett 84:24–32. https://doi.org/10.1785/0220110135
2. Minoura K, Imamura F, Sugawara D, Kono Y, Iwashita T (2001) The 869 Jogan tsunami deposit and recurrence interval of large-scale tsunami on the Pacific coast of Northeast Japan. J Nat Dis Sci 23:83–88
3. Silver N (2013) The signal and the noise: the art and science of prediction. Penguin, London

Part I

The Journey

2

First Lessons in Global Change

2.1 Youth Adventures

Several experiences with extreme weather were my first exposure to possible global change while growing up in the Minnesota River valley within tornado alley of North America. The first was with lightning on a summer day in 1946. A severe thunderstorm was taking place in North Mankato, Minnesota. Our whole family was eating lunch together in the kitchen. Lightning struck our house radio aerial and a fireball came down through the kitchen. Of course we were all terrified and ran out of the room to the living room, however at that time lightning again struck and split the tree in our front yard. When assessing the damage, we found that a fist-sized hole had been burned in the back steel plate of the stove and the frying pan on the stove had a hole melted through it that had sealed up. My sister still uses this pan today.

Do not believe that lightning can only strike once because I had many other experiences with lightning strikes. The second even more frightening experience occurred while my father, brother and I were backpacking in the northern Minnesota canoe country. We had portaged through two lakes and were standing on a large outcrop of granite on a gray day without rain. Suddenly, a bolt of lightning grounded on the granite block that we were standing on. The explosion of the lightning strike blew us apart. I looked up from lying on the ground to see my father and brother also lying on the ground about 3 m (10 feet) away from me. I thought they were dead, but then they too woke up and we started shaking together. We all had numb

C. H. Nelson, *Witness To A Changing Earth*,
https://doi.org/10.1007/978-3-030-71811-4_2

legs for the rest of the day and my brother's steel brace for polio was very hot, but did not burn him. Two later times in my life I was on planes that were struck by lightning, but there were no problems with the plane.

A second encounter with the extreme forces of nature also occurred during my early youth. During another severe storm, a tornado passed within a block of our house. We could hear it roar by, but we had no damage other than destroying my favorite woods to play in. Unfortunately, the tornado skipped to the other side of town and nearly 30 people were killed. I still remember seeing truckloads of injured people being carried through the streets to hospitals. Fortunately, the science of meteorology has advanced greatly and now accurate storm forecasts are made so that severe weather can be traced almost block by block through a town.

Unfortunately, these advances and the importance of science do not seem to have been recognized by some politicians. During the 1994 contract with America by the USA Congress, the head of a science committee wanted to eliminate the US Weather Bureau, which of course provides weather information for everyone including news outlets. When asked by a reporter how he would get the weather reports, he replied that he would just turn on the TV. Even in 2019 a non-scientist was nominated to head the US Weather Bureau and he wanted to privatize it. Privatization will greatly increase costs to taxpayers and cause a loss of scientific expertise that has taken decades to develop. This points out the severe ongoing problem with the lack of scientific knowledge or deliberate lying about scientific facts by some politicians, such as those who deny human-caused climate change.

From my own experience with the aforementioned extreme storms and the following extreme weather flood event, I have great empathy for what the global population is now facing. In 1950 the Minnesota River flooded the Mankato and North Mankato area where I lived. The river rose rapidly and the efforts to complete levees ahead of the rising river finally failed. The river broke through the levees and followed an old riverbed pathway in front of our house. My father rescued all of our family that he could fit in the car, and the water was rising to the car floorboard when he drove away. However, he had to leave me behind, so I began to carry things up from the basement. Soon rescue workers came by in a boat because the water now was 2–3 feet deep in front of our house and they insisted everybody had to leave. I was taken to the National Guard Armory and joined several thousand other people that had been rescued. My father did not know what had happened to me and spent much of the night trying to find me, which he did much to his relief early the next morning.

This flood was estimated to be a 1000-year flood (i.e. one expected every 1000 years) and kept our family out of our house for three months. After this flood, a much higher river levee was built. Several years later a similar flood took place along the Minnesota River. Fortunately, the levees held in my town of North Mankato, but this was a lesson that a greater number of major floods were taking place along the Minnesota River and its tributaries. Downstream from the Minnesota River, in the following years severe floods also took place along the entire Mississippi River. This increase in major floods from the Minnesota and Mississippi River system may have been a sign of a new pattern of more severe flooding (e.g. Kunkle et al. [15]).

Was my experience of the Minnesota River flood in 1950 a beginning sign of new weather patterns from human-caused global warming, because significant temperatures increases began after 1920 [12]? If you are not familiar with the question of global warming see the detailed explanation in Chap. 5. Each of my early experiences of extreme weather floods, lightning strikes and tornados may or may not have been beginning evidence for global warming. Scientists need to observe long-term multiple patterns of climate change to prove that global warming is part of the cause for the increased intensity of extreme weather events. However, the evidence for more and increased intensity of extreme weather events is building, particularly since the 1990s (see Chap. 5 Sect. 5.7).

Another adventurous experience of my childhood was occasional visits and then spending some summers at my father's original homestead farm near Spring Grove in southeastern Minnesota. My first visits to the farm began during World War II when the farm still had kerosene lanterns and an outdoor toilet, which was not enjoyable to use during the minus 32 °C (−25 °F) weather of a Minnesota winter. However these experiences on the farm gave me a feeling of what it was like to live close to nature like the majority of Americans did during the 1800s. For example, in the late 1940s and early 1950s we still used the McCormick reaper like the 1800s. After the reaper bundled the grain, I would have to take the grain bundles and make shocks. Grabbing these bundles full of thistles cut and stung my sweaty arms. However, it was great fun to be part of a threshing bee where the bundles were taken on a hay wagon to the thrashing machine. The thrashing of grain required a community of around 20 people from neighboring farms. Now when only about one percent of Americans live on farms, there is a loss of this community spirit and a close connection with nature.

A valid question is whether we have gained with the progress of agribusiness. Certainly, we have been able to feed the increase of five billion humans in my lifetime, but this has come at a cost. We are losing topsoil at a much greater rate than is renewable because of erosion, too much fertilizing, and spraying clouds of pesticides [34]. We no longer have chickens, pigs and cattle grazing naturally in pastures, in contrast to now being caged in buildings or feed lots. Farmers sit alone in their combines rather than working together as a community with their neighbors. Previously, family farms could determine their own destiny, and although subject to market fluctuations, were not feeling out of control by the whims of agribusiness and globalization. A return to the 1970's small is beautiful family farm approach could give many a feeling of more domination over their lives and would help preserve topsoil as well as providing healthier livestock and food sources.

I was lucky to have my farm experience and it proved helpful for my later career in oceanography. At sea, you need farm type experience and resourcefulness to solve problems when you are isolated on a ship in the ocean. It was interesting to find most of my scientific colleagues at the United States Geological Survey (USGS) were of Midwestern Scandinavian and rural backgrounds. In our USGS marine geology group we had two Larson's, Olson, Peterson, Carlson, Johnson, Nilsen and Nelson, as well as other Norwegians like Eittreim and Kvenvolden. I was never sure whether it was the mid-western farm backgrounds, or that we were naive and didn't realize how difficult it was to work seasick on a rolling ship. We did not have many surfer dudes; probably because they realized it was much more enjoyable to sit on the beach and look at the ocean rather than trying to work in storms at sea.

2.2   Beginning Biological Lessons

I had the good luck to have probably one of the best high school biology teachers in America in the 1950's and he won many teaching awards. Bob Hanlon at my Mankato, Minnesota high school was a pied piper who engaged even the most jaded student. He took students for bird watching in the early morning before classes started. He took some of us to South Dakota and northern Minnesota during school vacations for Audubon bird counts. For the Minnesota Audubon Christmas bird count, we camped out at minus 33 °C (−28 °F), but saw the rare Arctic snowy owl! In the summer after my high school senior year, Mr. Hanlon also took several students and me for a trip throughout Mexico. This provided the opportunity to expand my small town view of the world, observe other environments and see more human impacts on the environment.

One of the ways that Mr. Hanlon engaged students was to have each student report on the number of pheasants that they had shot during their hunting that week. Year by year the total number pheasants dwindled because the farmers were draining their sloughs, which were the prime habitat for pheasants to live in. When I went to my 50th high school reunion, I asked a friend how pheasant hunting was around Mankato and he said there were almost no pheasants left to hunt and you had to go to hunt in South Dakota. When I went to the 60th high school reunion, the Governor of Minnesota was in Mankato to start the pheasant-hunting season, but I am not sure why when there were almost none to hunt.

The fundamental lesson about global change that we learned in Bob Hanlon's advanced biology class was that it takes only one limiting factor, such as a lack of habitat for pheasants, to effect the population of a biological species. The increasing disconnect of humans from natural processes has resulted in a failure to remember this important biological control on any species. Any one of many limiting factor can cause a change and reduction in the population of the species. An analogy for we humans is that we may have all our organs functioning well, but if only one organ like our heart begins failing, this becomes a limiting factor for our life. The same is true for the habitat that the human population exists in. Consequently, if there is not enough water for the human population this will limit the size of the population, or if there is not enough farmland to grow food, the size of the population will be limited.

As a biological species, we humans face a number of limiting factors such as enough unpolluted air to breathe, clean water to drink and land to grow crops for food. All of these limiting factors are increasingly affected by human-caused global change. For example, when lecturing on an ecological tour to Antarctica in 1994, on dark rainy days everyone got badly sun burned and children on South Georgia Island had to go to school using umbrellas because of the human-caused ozone hole over the South Pole region (https://eartho bservatory.nasa.gov/world-of-change/Ozone). In another example, many of the USA water supplies contain prescription medications from humans and growth hormones from animals (e.g. [14]). Also antibiotics were present in 65% of the rivers from 72 countries that researchers from the University of York examined in 2019. In addition, toxic heavy metals have polluted areas of Chinese cropland, and global warming plus overgrazing by farm animals have destroyed areas of African cropland. The bottom line is that the Earth will survive our human-caused limiting factors, but will the human species? I think the human species can survive if we immediately start planning for a sustainable future.

A second important ecological lesson learned in high school is the carrying capacity of the environment (Fig. 2.1). This carrying capacity is determined by the resources consumed (e.g. water used, nutrients of soils depleted, land deforested etc.) and by the capacity to absorb wastes (e.g. contaminated water, toxic mining wastes etc.). The problem for any population, like the human species, is that if there is a limiting resource factor (e.g. polluted air or water, limited food), the population will decline or become extinct if the carrying capacity of the environment cannot sustain that resource. For humans it is important to realize that since 1950 there has been an exponential growth of population from 2.5 to 7.8 billion people and use of multiple resources (wikipedia.org/wiki/World_population).

There are many examples that the carrying capacity of the earth is threatened for many species (Fig. 2.1). The most recent global analysis estimates that the present species extinction rate is 10 to 100X compared to rates for the past 10 million years (United Nations IPBES [38]. For example, during the past 50 years, half the world's vertebrate species have become extinct (e.g. [5]. We do not want our human vertebrate population to suffer the same fate of decline or extinction because we have not planned for a sustainable use of the finite limited resources and carrying capacity of our planet. And do not

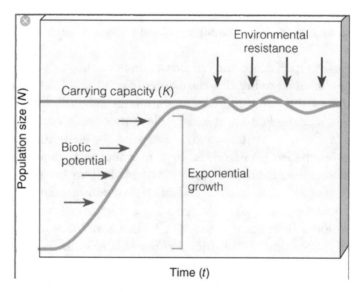

Fig. 2.1 Schematic diagrams showing human exponential population growth versus time with the eventual result that carrying capacity of environmental resources will limit population growth. It is estimated that the carrying capacity for the earth's resources is about 10 billion humans and this will be reached by the end of the twenty-first century [45]. Figure source is https://en.wikipedia.org/wiki/Carrying_cap acity

believe any suggestions that we can live on another planet, because we would rapidly deplete all our remaining resources of earth trying to move to another planet. This pie in the sky solution will not solve any of our problems caused by human global change.

The eighteenth century philosopher Thomas Malthus first noted the problem about the carrying capacity of the earth for the human species when he wrote: "The power of population is so superior to the power of the Earth to produce subsistence for man, that premature death must in some shape or other visit the human race" (Fig. 2.2) [16]. Many scientists now think the Earth has a maximum carrying capacity of 9 billion to 10 billion people. For example, the famous Harvard University sociobiologist Edward O. Wilson has suggested in his book "The Future of Life" (Knopf 2002) that the present 3.5 billion acres of farmland would support about 10 billion people. If projections of the United Nations Population Division are correct, the human population will reach the Earth's carrying capacity of either 9 billion by 2050, or 10 billion by 2100 [39]. This assumes that the Earth continues to have the present 3.5 billion acres of farmland, which is under considerable stress from global change (e.g. see Chaps. 5 and 6 Sect. 6.6).

Fortunately, we may be spared from entering the end-times phase of overpopulation and starvation envisioned by Malthus and Wilson because somewhere between 2050 and 2100, scientists think we'll make a U-turn in population growth (Fig. 2.2) (Wolchover, Live Science on Twitter 2017). The UN estimates of global population trends show that families are getting

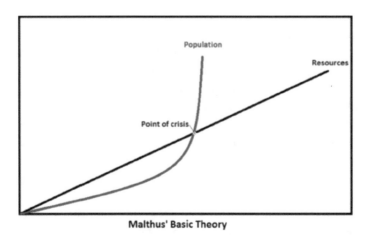

Malthus' Basic Theory

Fig. 2.2 Figure from Malthus [16] theory showing that eventually the exponential growth of population will reach a point of crisis when the number of humans exceeds the carrying capacity of the earth's resources. Figure source is https://study.com/aca demy/lesson/malthusian-theory-of-population-growth

smaller. "Empirical data from 230 countries since 1950 shows that the great majority have fertility declines," said Gerhard Heilig, chief of population estimates and projections section at the UN. Globally, the fertility rate is falling to the replacement level of 2.1 children per woman, the rate at which children replace their parents. If the global fertility rate does indeed reach replacement level by the end of the twenty-first century, then the human population will stabilize between 9 and 10 billion. As far as Earth's carrying capacity for humans is concerned, we'll have gone about as far as we can go, but no farther assuming present global conditions.

2.3   Broadening My Horizons

At Carleton College in Minnesota, I continued my studies in biology and then concentrated on earth science. Specific studies and field trips in ecology broadened my understanding of limiting factors and carrying capacity in different environments like forests, rivers, lakes, and oceans. Earth science studies increased my knowledge of limited mineral resources and the natural catastrophes that can cause global change.

One of my first lessons about a natural geologic catastrophe occurred during my career at Carleton when I was a geology major and fortunate to encounter the second important mentor in my life, Dr. Eiler Henrickson. He specialized in mineral deposits and in the summer of 1958 he worked for a mining exploration company in Alaska. They established a base camp high on the mountainside of Latuya Bay. Fortunately he was offshore in a boat and the tides prevented him from returning to the base camp because that night a great earthquake occurred on the Fairweather Fault under Latuya Bay and created the highest tsunami in historical times (525 m) (1700 feet) [19]. This tsunami reached the 1700 feet up the mountainside and completely washed away all the forest and Eiler's base camp. The tsunami killed several fishermen, however, one fishermen's boat miraculously rode out the 300-foot wave at the Bay entrance and Eiler rescued him. Little did I realize that many decades later, I would study the effects of several tsunamis in the Ionian Sea that killed tens of thousands of Italians over the last several hundred years [35].

My geology classmate Dr. Walter Alvarez shows another example of how my classmates and my undergraduate experience at Carleton trained me to understand natural events with global implications. Walter and his father Dr. Luis Alvarez, a Nobel Prize winner, developed the theory of dinosaur extinction caused by a meteorite hitting the earth [2]. This shows that global changes from natural catastrophes as well as human impacts, such as on pheasant habitat, can help eliminate species. However, there seem to have been only several of these types of natural catastrophes throughout the several billion-year history of the earth. In contrast, from our worldview as earth scientists, we can see that a number of human-caused global changes during just the past half-century are depleting resources, limiting the earths carrying capacity for living species and causing species extinction rates similar to those observed from natural catastrophes over millions of years (Fig. 2.1) (see detailed discussion in Chaps. 5 and 6) (e.g. [5]).

Two other Carleton alumni from my era, Drs. Donella and Dennis Meadows in their 1972 book "Limits to Growth" caught the world's attention about the sustainable use of resources [17]. Their computer models originally predicted that many mineral resources would be depleted by the year 2000. However, geological exploration to find new sources of these minerals and substitution of plastics for metals resulted in the fact that most of these resources had not been depleted by 2000. This illustrates that humans have the ability to innovate changes for sustainability, although the 2004 revised models of the Meadows still estimate that many mineral resources will be depleted in around 100 years [18].

The new Meadows book with a 30-year update of their 1972 book on limits to growth provides new insights about the basic carrying capacity of the earth to sustain the global population [18]. They postulate that there are two possibilities for the carrying capacity of the earth. The first possibility is that there will be no change in the use of our limited resources and that the human population will crash. Dr. Jared Diamond's book "Collapse" provides many examples of societies that depleted one or another of their limited resources and the society collapsed. The second possibility is that there will be a correction to a sustainable use of these limited resources and the carrying capacity of the earth will continue to support the human population. The Meadows analysis indicates that in the 1980's the human population began using a number of limited resources at a greater rate than can be regenerated. Thus, at the rate of resource use in the 1980s, the human population began overshooting the carrying capacity of the earth.

Since the year 2000, this overshooting has only accelerated with the important economic development in China and India. One manifestation of this development is that oil prices increased to over $100 dollars a barrel. Again, there is evidence that humans have the ability to innovate global changes for sustainability, because fracking has resulted in new sources of petroleum. As a result of the new petroleum resources and slower growth in China, the petroleum prices began collapsing in 2015.

Unfortunately, these short-term price changes do not solve the problem of long-term sustainability of energy. Also the fracking processes create other global change problems such as the introduction of the green house gas methane (CH_4), which is more potent than carbon dioxide (CO_2) for global warming. Studies need to determine whether the earth has the capacity to absorb the toxic wastes from fracking, and whether other environmental problems can be solved such as ground water contamination and initiation of earthquakes generated by the fluid injection processes of fracking. The long term solution for a sustainable energy future will need to be renewable energy from a mix of sources such as hydroelectric, wind mills, and solar panels. Northern European countries like Denmark, Scotland and Spain in southern Europe have shown already that where there is the political will and investment in infrastructure, they can provide up to 100% sustainable energy from renewable sources (https://www.zmescience.com/ecology/environmental-issues/denmark-scotland-renewable-energy-environment-06062012/).

Another one of the earliest examples of global political will for the environment was associated with my undergraduate experience at Carlton College. The president of Carleton at that time was Dr. Lawrence M. Gould. Before becoming president of Carleton, he was famous as an Antarctic explorer, because was second in command for the Admiral Byrd expedition to Antarctica from 1928 to 1930 [11]. After his experience from the Antarctic expedition, Dr. Gould became a famous speaker in the early 1930's. His interest continued for Antarctica and in 1957, while I was a student at Carleton, he became the leader of the International Geophysical Year. This was a global initiative for scientific expeditions to Antarctica by many countries (https://en.wikipedia.org/wiki/Laurence_McKinley_Gould).

Following this International Geophysical Year, the Antarctic Treaty was formed to protect Antarctica for peace and science and has not permitted any global exploitation for resources. Thus, thanks to the leadership of Dr. Gould and international co-operation, Antarctica has had a protected environment since the international treaty. The only development has been for scientific laboratories by different treaty members. Unfortunately, this has not been true for the marine offshore areas where whaling continues and affects the whale populations. As a tribute to his leadership, a USA Antarctic research ship is called the Lawrence M. Gould.

Every year the Antarctic Treaty members meet to continue and make any revisions to the Treaty. These meetings are similar to the United Nations where all of the countries sit around a large circle with each country at its table to make the treaty negotiations. Because my wife, Dr. Carlota Escutia, presented the only scientific talk at one of these Treaty meetings, I was able to attend this meeting in Brasilia (Brazil) in 2014. The significance of the Treaty is that many of the world's countries work together to protect the environment of Antarctica in a global effort. This is the type of international activity that should be used in many areas to protect our Earth from global change. The 2015 Paris Climate Agreement on global warming is another example of an international effort to maintain a sustainable planet for humans. Unfortunately, the USA became the world's only nation that has dropped out of the Paris Agreement, but fortunately has now rejoined.

Another important accomplishment of Dr. Gould is that he developed the outstanding geology department at Carleton College. I became a combined geology and biology major at Carlton, which became a turning point for me to become a natural scientist studying the earth's environment. As a result, I have been lucky enough to conduct worldwide research, which has given me a wide range of insights to observe global change. One of these insights is that in 1994, while lecturing on an ecological tour of Antarctica, I was able to observe the results of Dr. Gould's vision for the Antarctic Treaty. When we made 25 Zodiac landings on the tour, you could walk among pristine colonies of millions of penguins, elephant seals, bearded seals, albatross etc., where these animals and birds had no fear of humans resulting from interference such as hunting. You also could observe all the natural biological interactions such as hawks swooping in to steal penguin eggs, brutal fights between male bearded seals or elephant seals to protect their female harems.

2.4   Crater Lake Turning Point

A few days after graduating from Carleton College in 1959, I headed for Crater Lake National Park, Oregon (CLNP), which became a turning point in my life and led to continued Crater Lake research throughout my scientific career (Fig. 2.3). Perhaps growing up in Minnesota surrounded by many lakes inspired me later to study Minnesota lakes, Crater Lake, Oregon and Lake Baikal, Russia (e.g. [22, 25]. Crater Lake became a theme of my geological journey because of my four summers as a Seasonal Ranger Naturalist, master's degree research at the Lake, discoveries of Crater Lake ash in the deep sea floor off Oregon during my PhD research, and USGS geological hazard studies on the lake in from 1979 into the 1980s [32, 26, 24]. Now during my retirement, Crater Lake ash in the deep sea off the Pacific Northwest has become a key to determining the great earthquake history and hazards of the Cascadia Subduction Zone in the Pacific Northwest (e.g. [29]).

Fig. 2.3 **a** Crater Lake research boat with Hans Nelson ready to lower a grab sampler to obtain sediment on the bottom of Crater Lake for his MS thesis research in 1960. **b** Glass jar with grab sample of sediment in 1960 held by Hans Nelson (e.g. no plastic containers at that time). This photo is shown in the pioneer scientists exhibit at the Sinnott Memorial Overlook in Crater Lake National Park. The caption reads: During the late 1950s Hans Nelson used coring devices and small dredges to obtain sediment and rock samples from the floor of Crater Lake. His studies of pollen and plankton found in these sediments set the stage for our current understanding of the link between the lake's physical and biological environments. Later as a marine geologist with the U.S. Geological Survey, Nelson provided the first detailed description of the lake basin and the distribution of sediments found there. His research laid the foundation for later studies that located thermal areas in Crater Lake. Photo source is U.S. National Park Service

The first significant global change lesson that I learned with my experience at CLNP has been the value the US National Park Service provides by preserving natural areas. After almost 60 years of working and visiting CLNP, it is one of the few places on earth I have visited where significant global change from human activities has not occurred, except for the increased number of dead trees from drought and forest fires. The National Park system in the United States has set the global standard and numerous other USA wilderness and marine sanctuary areas are a defense against global change. The National Park Service has a constant battle between preservation and use. Large areas of National Parks are preserved, but some areas have major impacts from visitors. However, the Park visitation is extremely valuable because it is a way that humans can connect with nature.

Human connections with nature in the developed world are becoming increasingly important as people live more and more in cities and the virtual world of computers rather than the natural world as most people did a century ago. All of the modern technology and separation from the natural world leave people with the impression that technology can solve all our global change problems. When the increasingly intense storms occur from global warming, and natural catastrophes take place like the recent Sumatran and Japanese earthquakes and tsunamis, humans are jolted back to reality that we live in a natural world that still controls our destiny. One example of this disconnect caused by the virtual world is that National Park yearly visitation decreased during the early millenium rather than generally increasing yearly as it has done again recently (https://www.nps.gov/aboutus/visitation-numbers.htm). The National Park Service attributed this temporary visitation decline to people spending more time in the computer virtual world instead of visiting the natural world of parks.

Another important preservation activity of the USA National Park Service has been the administration of offshore marine sanctuaries or Marine Protected Areas since their establishment in 2000. Protected areas are necessary because the world fishing fleet has the overcapacity to destroy the entire wild marine fisheries (United Nations [38], IPBES). One third of the marine fisheries, such as the Grand Banks cod, already are depleted, and the entire marine wild fisheries may disappear in about 50 years [39]. At present fishing rates, this global change may take place in one lifetime. The only protection against this change is maintaining large areas of marine sanctuaries where threatened species can regenerate and fisheries can recover. However this requires political will, regulations and enforcement to prevent overfishing.

The USA has taken the steps to develop well-managed, sustainable marine fisheries for our marine territory. The USA has set an example for most of the rest of the world for a way to prevent this human-caused global change to marine fisheries. Unfortunately, for the first time the USA president has removed some areas of the USA National Park System from protection in 2019, which is certainly a backward step for global change problems.

Fortunately, a progressive activity in the USA, like the development of Marine Protected Areas, is that some of the critical estuarine nurturing grounds for fisheries are being restored to natural habitats. Anyone landing at San Francisco airport has flown over the brightly colored salt evaporation ponds. These ponds have been constructed by damming off much of San Francisco Bay, which is the largest estuary in western North America. This, along with all the Bay area development, such as sewage outfalls into the Bay, has destroyed most of the Bay habitat. For example, Jack London described the previous natural habitat with extensive oyster beds and abundant sea bass fishing in the Bay, which no longer exist. However, some of these salt ponds are now being destroyed so that Bay habitat can regenerate. This is a complicated battle for restoration, because introduction of the exotic Chinese zebra clams from ship bottoms has completely altered the biology of the Bay. This is yet another example of global changes from our interconnected world.

My Masters degree lake studies at the Limnology Institute at University of Minnesota pointed out many other examples of human global change from our interconnected world. At the University of Minnesota and elsewhere in the early 1960s, limnologists (lake scientists) found other evidence of how human activities affected our global environment. It was known that coal fired power plants introduced CO_2 into the atmosphere. However, these new studies in Minnesota and other areas showed that coal-fired plants introduced other pollutants into the air. Those that caused most problems for lakes were hydrogen sulfide and mercury. With rainfall the hydrogen sulfide became sulfuric acid. The acid rain then made the lakes more acid, which significantly affected the biology of the lakes since with increasing acidity many organisms could not survive (e.g. [4]).

Mercury from the air pollution entered the lakes and microscopic organisms of the lake ate the mercury that collected in the lake water (e.g. [37]). Then through the process of biological concentration up the food chain, as small fish ate the microscopic organisms and then the larger fish ate the smaller fish, the mercury concentrated in dangerous amounts in the largest fish. Mercury pollution in fish remains a problem to this day. When fish were tested from 300 streams in the USA, all of them had mercury contamination and 27% had mercury levels high enough to exceed what the EPA considers safe enough for the average fish eater consuming fish twice a week [47].

Studying lakes in Minnesota was an adventure because you not only used boats during the summer, but you also could do studies much cheaper by drilling through the ice in the winter. The sampling team would drive out onto the ice and then drill a hole similar to that done for ice fishing. Then in this minus 32 °C (−25 °F) weather you would lower down core barrels to take samples of the lake bottom mud. These barrels took a plastic tube full of mud, or core, that was then taken back to the laboratory to open and study the sediment layers going from youngest at the top to oldest at the bottom.

The sediment layers are like pages of a book showing the history of the lake. The pollen grains provide the history of the forest. The pollen history shows how the original hardwood and coniferous forests changed to grain and cornfields when farming began. Layers of wood ash and burned wood fragments indicate when forest fires occurred and then the history of natural fires can be compared to the increased number of forest fires that result from droughts and global warming. The chemistry of the lake sediment can show the lake becoming more acid and polluted with mercury from burning coal and lead from cars burning leaded fuel. When this lead pollution was found in lakes, on land, in the air and then globally in the ocean, the world community acted and removed lead from gasoline to reduce lead pollution.

The sediment layers can show the history of other natural events like severe storms and floods plus how often they have been taking place and if this history has changed related to changing climate. Recently, my colleagues have even been able to use lake sediment layers to determine the history of catastrophic great earthquakes (Mw = or >8) that have global effects. When strong shaking from earthquakes causes sediment to flow into the center of lakes, we can use these coarser silt and sand layers to identify the earthquake events (e.g. [24]). By determining the recurrence time between these earthquake sediment layers, scientists can see how frequent these earthquakes with global impacts, such as ocean wide tsunamis, have been on a fault system.

Colleagues have used this method in Cascade mountain lakes to help determine the hazards and history of the Cascadia Subduction Zone earthquakes for thousands of years [20]. In Chilean lakes the 1960 earthquake, the world's largest historic earthquake can be identified in lake sediment and compared with many earlier Chilean earthquakes to determine the global hazards of the great earthquakes on this fault [21]. We have witnessed how the Chilean, 1960, Alaskan 1964, Sumatran, 2004 and Japanese 2011 great earthquake tsunamis crossing oceans have had worldwide effects.

Similar to the natural global change of great earthquakes, catastrophic volcanic eruptions such as Mount Mazama also have had global effects of changing the climate, as well as forming the caldera that contains Crater Lake, Oregon (Fig. 2.4). My MS thesis and later studies on Crater Lake sediment layers helped to define the volcanic history of Mount Mazama, and assess the potential geologic hazards of future eruptions [22, 32, 26, 24]. The sediment and rock samples that I collected from the lake floor showed that after the cataclysmic eruption filled the atmosphere with volcanic ash and cooled the world climate, several other volcanoes erupted on the caldera floor as it filled with water. I took the first samples from the submerged cones and found out that Merriam Cone was a cinder cone like Wizard Island and

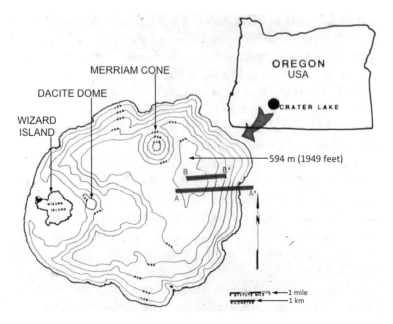

Fig. 2.4 Map showing the location of Crater Lake in the Cascade Mountains of Oregon and the water depth of the lake floor. Seismic profile location for Fig. 3.2 is shown by the line B – B' and core transect location for Fig. 2.5 is shown by line A—A'. The Mount Mazama eruption and collapse to form Crater Lake occurred about 7600 calendar years ago [42]. The lake is about 10 km in diameter and 600 m (1950 feet) deep. After the collapse and while the lake was filling volcanic eruptions continued to form the submerged Merriam Cone in the north, Wizard Island rising above the water on the west and the submerged rhyodacite dome just east of Wizard Island. Figure source is Nelson et al. [32]

the dome east of the Island was dacite volcanic glass (Fig. 2.4) [22]. During later studies in the 1980s I determined that the submerged volcanic ryodacite dome erupted about 4,000 years ago and Mount Mazama has not erupted since (Fig. 2.4) [24]. Additional studies that I will describe later suggest that there are no indications in Crater Lake of an imminent volcanic eruption of Mount Mazama.

Similar to my Crater Lake studies to define the lake floor biology, sediment history and natural global hazards, the small lake environments are the first and best environments to alert us to human-caused global change [22, 24]. Lake Tahoe on the border of Nevada and California in USA has been a classic example to show the many global changes to our environment. The first problem detected in Lake Tahoe was when the pristine clear water began to develop algae growth from increasing biologic productivity. This increase in productivity resulted from rapid population development around the lake and the dumping of raw sewage into the lake (e.g. [10]). The sewage increased the nitrogen, phosphorus and other nutrients, which resulted in unusual productivity for a clear mountain lake. To save Lake Tahoe, the world's first tertiary treatment sewage plant was developed so that sewage could be treated and the resulting water was drinkable. This plant became world famous and was visited by governments from many countries as an example for a solution to lake and other water pollution.

Unfortunately, this was not the end of the environmental problems for Lake Tahoe. After the lake began to recover from the sewage pollution, again unusual biological productivity began. It was discovered that people were fertilizing lawns around the lake and this fertilizer of nitrogen and other nutrients drained into the lake to increase the productivity of the lake (e.g. [3]). The use of plant fertilizers was prohibited and once again Lake Tahoe began to recover its water clarity. Then, as scientists had predicted, the lakes around the world began to suffer pollution problems from the high compression gasoline and diesel engines. These engines create more nitrous oxide in the atmosphere and then the rain brings this nitrogen into Lake Tahoe and other lakes to artificially fertilize them (e.g. [10]). Once again Lake Tahoe encountered algal blooms and began to lose its pristine characteristics because the normal Nitrogen to Phosphorous levels of 1:1 have reached 40:1 (Charles Goldman, University of California Davis, personal communication, 2018). Because of this latest nitrate contamination, the lake water clarity as measured by the secchi disk has declined 40% from the 1968 depth of 31 m (102) feet visibility to 20 m (64 feet in 2016) (Nevada Sagebrush, Nov. 24, 2017).

I witnessed an interesting analog to Lake Tahoe when I was studying Lake Baikal [9]. This Russian lake is the world's oldest, deepest and largest lake by volume, because it contains 20% of the world's fresh water. Lake Baikal, like Lake Tahoe, provides an example of unusual pollution for a cold pristine mountain lake with clear water in most locations. Unfortunately on the shoreline of the South Basin of Lake Baikal, the world's largest paper mill discharged wastes into the lake until it was shut down in 2013 (https://en.wikipedia.org/wiki/Baykalsk). As a result of the paper mill pollution, in 1992 the South Basin was the only location in Baikal where I observed wide areas of floating masses of algae.

As you can see, this battle against global change continues in lakes worldwide and you can see how many human activities are influencing our global environment. Even more disturbing, just as the smaller lake environments have been affected, we see many of this same changes taking place in the ocean that covers 70% of the Earth. Scientists now observe that the acid increase first found in lakes has now spread to the entire ocean (see Chap. 5, Sect. 5.3). The increasing acidity of the entire ocean is resulting from the increased CO_2 in the atmosphere that is caused by burning fossil fuels. Observations like this are why we scientists believe that our planet has entered a new geologic time, where human activity is now controlling the fate of the Earth's history.

In 1962 I undertook another type of study to assess the effects of global change from chemical pollution in rivers and estuaries. I analyzed living clam flesh for heavy metal pollution (e.g. toxic heavy metals such as mercury, arsenic, lead). New Jersey has a large complex of chemical factories and the Toms River drains through it. To determine the pollution effect of the river, I collected clams from the river because they filter out and concentrate the toxic metals in their bodies. I determined that there were elevated levels of toxic heavy metals, which showed an example of the global pollution that humans were causing. Many other scientists independently around the globe have now used the same method of studying clams to assess polluted aquatic environments.

2.5   Storm of a Century

There were many adventures while using boats to take samples from the bottom of the Crater Lake volcanic caldera. The lake has a maximum depth of 594 m (1949 feet) and is the 7th deepest lake in the world (Fig. 2.4). My first day taking samples on the lake in 1960, I was using a boat built in 1932, and a 1928 Dodge generator motor to power a winch with airplane cable wire (Fig. 2.3a). Prior to heading out my first day to sample, I had stumbled into my dorm room after a night at the bar and found two gentlemen in sleeping bags on the floor. One of them was Parke Snavely, who was later to become the first Branch Chief for Marine Geology in the USGS. The second gentleman was Dr. Dallas Peck, who was later to become the Director of the USGS. We rousted out at 6 AM and then had to hike a mile and a half down the 1200-foot high caldera wall to the lakeshore. We went to the deepest part of the lake and lowered the coring tube and weights to the bottom (Fig. 2.4). When we tried to pull up the corer from the bottom of the lake, the winch motor failed. We spent the rest of the day taking turns winding up the nearly 2000 feet of cable. Apparently, these USGS geologists were impressed by my persistence, because a few years later they hired me for a permanent job with the USGS, where I worked for 32 years. Also, a picture from my 1960 sampling on the lake now is found on the wall of the CLNP Sinnott Memorial Overlook to show me as one of the pioneer scientists at CLNP (Fig. 2.3b).

The 1962 Columbus Day storm in the Pacific Northwest of America was a much more dangerous adventure than the first day of Crater Lake sampling (https://en.wikipedia.org/wiki/Columbus_Day_Storm_of_1962). This storm occurred a few days after I moved permanently to Oregon for teaching at Portland State University and eventual PhD studies at Oregon State University. This storm may have been a several hundred-year event, because peak wind speeds were 270 km/hr (170 mph) and 30% of the Cascade Mountain forests of the Pacific Northwest were destroyed. As the storm began, I arrived at the school where my wife was teaching and the wind was so strong that I had great difficulty opening my car door. As my wife and I were standing outside under the school foyer, several other teachers assured us that such storms were a yearly event. We began to doubt this wisdom as we observed that the 30–60 m (100 to 200 foot) trees in the distance were being blown down like toothpicks. When the entire 5 m (15 foot) wide foyer above us was blown up into the air and crashed through the school roof behind us, the long-term Oregon residents were convinced that this was not a typical storm. We were without electricity for two weeks and my university lectures were prepared by candlelight. Was this Columbus Day storm a precursor of the increasing intensity of extreme storms during the past half-century, such as hurricanes Katrina, Harvey, and Iris, which appear to be intensified by human induced global warming?

2.6 The Deep Sea PhD

The biggest expansion of my global view began when I started my PhD studies in geological oceanography at Oregon State University in 1963. This was very exciting for me because I could apply all the sciences (physics, chemistry, biology, and geology) to one global environment that covers 70% of the earth. I was extremely lucky to begin at Oregon State as one of the first students in oceanography. The school had gathered a great group of professors who later spread throughout the United States as chairman of other marine science departments. Also during these early years of the 1960s, the funding and ship availability for marine studies were good thanks to President Kennedy. Because President Kennedy enjoyed sailing, he began a major program to increase oceanographic studies for the United States. Ten universities were selected to become centers for oceanographic studies and Oregon State was one of these new institutions. It was tragic that in my first fall of studies at Oregon State, my physical oceanography class was disrupted on November 1963 when President Kennedy was assassinated. Everything at the university came to a halt for several days; however President Kennedy's legacy has lived on in marine science studies.

My doctoral studies at Oregon State specialized in geological oceanography or marine geology as it also is called. When people ask about my work, they usually say, oh you do a lot of scuba diving. Some marine geologists have the enjoyable specialty of scuba diving and studying coral reefs. However, in the Pacific Ocean off Oregon or Alaska where I did many of my scientific studies, coral reefs were not an option. Consequently, I used the typical type of techniques for marine geology. These techniques for marine geologists consist of mainly two types, using sound sources emitted from ship equipment or coring samplers and other equipment that are dropped into the seafloor or lake floors to collect samples of the ocean bottom sediment (Figs. 2.3, 2.5, 2.7b).

The sound sources vary from high frequencies used by echo sounders to low frequency sound waves that penetrate far below the seafloor (Fig. 2.5). The sound frequencies vary from those that penetrate only a few meters below the seafloor (Fig. 2.6b) to low frequencies that penetrate kilometres below the seafloor (Fig. 2.6a). In the early days of marine geology to create the largest and lowest frequency sound waves, bundles of dynamite where lit and thrown off the ship. However, this dangerous technique was substituted by other sound sources such as giant multiple air guns that explode compressed air to make sound waves, although this noise also may be harmful to whales and even much smaller ocean life [48].

Similar to the advances in acoustic techniques (Fig. 2.5), there have been advances in sampling and studying lake and ocean bottom sediment. Although cores of bottom sediment remain important (Fig. 2.7b), present day techniques include heat and geotechnical probes, submersibles, television cameras, bottom cameras and remotely operated vehicles (ROVs), which can observe the in situ bottom conditions. In shipboard and onshore laboratories, in addition to the classical sediment sampling (Fig. 2.7b), cores can be scanned for physical properties (e.g. density), chemical composition, and CAT for detailed structures in the sediment layers.

My PhD studies focused on the surface and near-surface marine geology of Astoria Fan, a submarine fan deposited on the deep seafloor off the submarine Astoria Canyon and the Columbia River in Oregon, USA (Fig. 2.7a, 2.8). In the early days of my thesis work, we only had echo sounders. These echo sounders sent a single sound wave that went to the seafloor surface and bounced back to be recorded and show the water depth on a paper recorder (Fig. 2.5). As the ship traveled along, the echo sounder continuously sent out sound waves that were recorded. This created a line of the depths that could

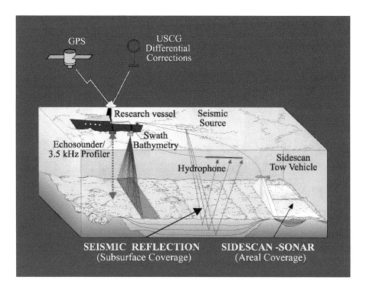

Fig. 2.5 Schematic diagram showing how marine geology utilizes different sound wave acoustic techniques including single echosounder waves for water depth, multiple echosounder waves for swath bathymetry of wide areas of seafloor depth, sidescan sonar scanning sound beams reflecting off the seafloor surface to show morphology, 3.5 kHz high frequency seismic reflection sound beams that reflect off subsurface geology near the seafloor (see Fig. 2.6b seismic profile example), and powerful low frequency seismic reflection sound beams that reflect off the deep subsurface geology below the seafloor (see Fig. 2.6a seismic profile example). All of these acoustic systems send sound waves that bounce off the seafloor surface, near-surface (e.g. Wing et al. [43], Levi et al. 2019) sediment layers or off deep subsurface layers. The returning sound wave signals are received by hydrophones on the ship bottom or towed behind the ship. Recorders process and print the signals so that the surface and subsurface geology can be viewed and interpreted (see Fig. 2.6 seismic profile examples). The ship is continually moving while employing these acoustic methods and thus accurate GPS (Global Position System) ship navigation must also be recorded and correlated with the time lines on the profiles to know the location of the interpreted geology. Figure source is the U. S. Geological Survey

be put on a sea floor map. These lines of depth could then be contoured to show the shape of the seafloor much like an onshore topographic map. During my career, great advances in echo sounding techniques have taken place. Now arrays of sound waves can be sent out from the ship to record wide swaths of depth (swath bathymetry) and shape of the seafloor (Figs. 2.5, 2.8). High frequencies of sound waves can be used so that high-resolution swaths of a few tens of meters can be taken to study small features such as from whale and walrus feeding on the seafloor. Other low frequencies can obtain a swath up to 70 km (42 miles) wide, so that large areas of the seafloor morphology

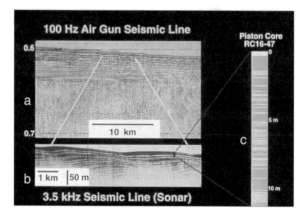

Fig. 2.6 a Example of a low resolution deep penetration 100 Hz seismic profile, **b** a high resolution 3.5 kHz profile over part of the same area as the 100 Hz profile in a, and **c** drawing of a piston core tube of seafloor sediment taken in the area of the 3.5 kHz profile in B. The yellow sand/silt layers in the piston core cause the dark layer reflections in the 3.5 kHz profile. See Fig. 2.7b for a photo of an actual piston core in which the sand/silt layers are being described. Figure source is the U. S. Geological Survey

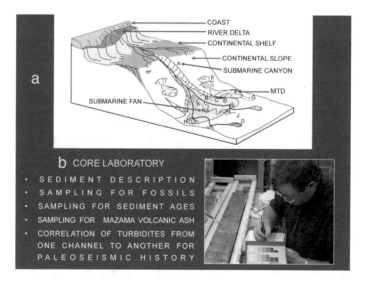

Fig. 2.7 a Schematic diagram of a source to sink continental margin showing the coast, river distributaries feeding a sediment delta, continental shelf, continental slope eroded by a submarine canyon (see A in figure) that feeds a submarine fan with channels and turbidite sand layers (see K in figure). MTD is a submarine land-slide deposit (see F in figure). Figure source Close (2010). **b** Description of shipboard core laboratory research and photo of my late friend Dr. Eugene Karabonov as he describes the deep-sea turbidite sand/silt layers generated by great earthquakes along the northern San Andreas Fault in California, USA. Figure source is Hans Nelson

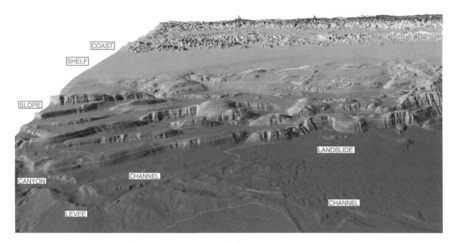

Fig. 2.8 Map showing the three dimensional rendition of swath bathemetry from the northern Oregon continental margin in Cascadia Basin.The figure has the coastal topographry integrated with the bathymetry and has labels for the continental shelf, continental slope, and Astoria Fan features of Astoria Canyon mouth, Astoria Channel pathway (channel below label), Astoria Channel levee (above label). About 50 km (30 miles) of the proximal Astoria Fan is shown and the proximal Astoria Channel is 5 km (3 miles) wide with 250 m (800 feet) of relief. A submarine land-slide at the base of the continental slope is labeled and the lower continental slope has an eroded scarp that is the source of the landslide. Also the slope exhibits the accretionary wedges caused by the Juan de Fuca tectonic plate that is sliding under the continental margin and causing Cascadia Subduction Zone earthquakes. Figure source is [44]

can be studied. This is valuable because the seafloor covers 70% of the earth and only a small portion has been mapped.

For nearsurface studies of the sea bottom, high frequencies of sound waves for seismic profiling are used to obtain details of the shallow sedimentary layers near the seafloor surface (Fig. 2.6b). As the ship travels, sound waves are continually sent out to penetrate below the seafloor and bounce off the subsurface sediment layers. The signatures of the returning sound waves are then recorded on paper recorders or computers in the ship laboratory. Differences in the sediment layers below the seafloor can then be observed to show composition of layers such as mud or sand and these can be confirmed in sediment cores (2.6b, c and 2.7b). Also geometries can the seen, such as sediment waves on the seafloor or buried below a seafloor (Fig. 2.6). By observing these different geometries and characteristics of the subsurface layers, some of the history of the seafloor can be obtained.

After sounding devices have identified seafloor characteristics like submarine channels, then sediment cores can be taken to find out the types of deposits in the channels (Figs. 2.7b, 2.8). For example, when the submarine channel floor deposits are cored, sand deposits are found just like river channels on land (Figs. 2.7b, 2.9b) [23]. The channel levees in the deep sea have the same deposits as the levees of river channels (Fig. 2.8). In this way, the processes and history of the deep seafloor can be obtained. For example, when looking at cores from my PhD thesis area, I discovered that the Mount Mazama ash from the Crater Lake volcano was in the sand layers of the submarine channels off Oregon (Fig. 2.9b). From these deep sea deposits, I was able determine that the ash from the Crater Lake eruption was eroded

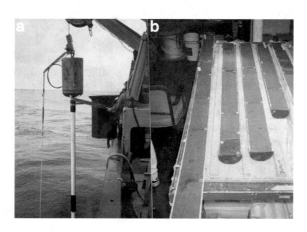

Fig. 2.9 **a** Photo of a piston corer prepared to be lowered by a ship winch cable. The cable just below the pulley has a trigger arm attached to it and the large corer top with several thousand pounds of lead weight is held by the trigger arm. The trigger arm extends to the left of the corer and has a cable at the end of the arm. The trigger arm cable has a weight at the bottom of the cable. When the entire corer and trigger arm assembly is dropped to the bottom of the sea, the trigger weight hits the sea bottom first and when the weight is taken off the trigger arm, the main piston corer is released and drops into the sea floor while a piston attached to the end of the ship winch cable is pulled up through the core barrel to help the corer penetrate the sea bottom sediment. The piston corer penetrates much deeper into the sea floor than a gravity corer just pushed into the sea bottom by a weight at the top. Below the piston corer head with the lead weights is the white core barrel that is 3 m (10 feet) long in the section that is visible. If the trigger arm pre-trips in the sea and releases the corer, it falls with great velocity until it hits the end of the ship cable, which usually breaks. See discussion in Chap. 3 (Sect. 3.6) and Chap. 4 (Sect. 4.6) where pre-tripping of a piston corer has happened on my scientific research cruises. **b** Piston core barrels cut in half to see the sediment layers. The darker gray layers are sand turbidites that are marked with white plastic buttons. In these cores the turbidites are seismo-turbidites generated by Cascadia Subduction Zone earthquakes (see discussion in Chap. 4, Sect. 4.13). Note in the left core how evenly spaced the seismo-turbidite layers are because of the cyclic triggering of earthquakes about every 500 years (see Fig. 4.21 defining this earthquake recurrence interval). Photo source is Julia Gutierrez Pastor

off the land by rivers, then carried down submarine canyons by currents and finally deposited in submarine channels of the deep seafloor [31].

The presence of the Mazama ash was a key to understanding the depositional processes of the deep-sea sand layers. At the time of my thesis it was known that submarine landslides of sand and mud in submarine canyons could mix with water and accelerate down slope by gravity because of this denser mixture of sediment and water. These density current flows were called turbidity currents, which funneled down the submarine canyons, and deposited turbidite sand layers in channels on the deep seafloor (Figs. 2.7a, 2.8, 2.9b). Because of Mazama ash contained within the turbidite sand layers, I could trace the history of deposition of a single turbidity current event through Astoria Canyon and Channel and observe thicker sand deposits on their floors and correlative thinner sands above in the adjacent channel levees (Fig. 2.8) [31, 23].

My observation of Mazama ash in Astoria Fan sediment layers was important at the time, but became even more important many years later because we used the age of the Mazama ash to determine the earthquake history for the Cascadia Subduction Zone (Fig. 2.9b) (e.g. [1, 29]). By determining the great earthquake history, we defined the earthquake hazards that the Pacific Northwest needs to plan for and the global hazards of these types of the world's strongest earthquakes (Fig. 2.7b) (e.g. [30, 36].

The background for these later earthquake studies followed from my PhD research on Astoria submarine fan (Fig. 2.8). These submarine fans are like alluvial fans on land and have similar channels, although the scale can be much larger than the alluvial fans on land (Fig. 2.8) [27]. Submarine fans may be a few kilometers in diameter off small rivers, and reach up to 2500 km (1550 miles) diameter off the Ganges Brahmaputra River in the deep Bay of Bengal off Asia. Astoria Fan, however, is about 300 km (180 miles) in diameter (Fig. 2.8). My study was the first detailed study of a submarine fan system that took transits across the fan and channels, plus looked at changes from proximal to distal locations on the fan [23].

It was lucky that I picked Astoria Fan as my PhD thesis topic because later it was found that ancient subsurface submarine fans are good petroleum systems (e.g. [33, 29]). During the last few decades, some of the world's largest petroleum deposits have been found in buried subsurface ancient fans. After my initial study of Astoria Fan, there now have been thousands of other studies of modern seafloor, ancient outcrop and subsurface submarine fans because of their importance for petroleum resources. As a result, since my thesis I have continued to conduct studies of other submarine fan systems around the world. My choice for a doctoral thesis topic turned out to be fortunate for global resource as well as earthquake hazard studies.

Before beginning my thesis studies I had to learn the complicated process of becoming the chief scientist for a scientific cruise. The first task before having a scientific cruise is to obtain the money for a cruise. This has been extremely difficult throughout history from the time of Columbus to the present. The expensive costs of ships must be funded and nowdays this takes a minimum of a few thousand dollars to a maximum of millions of dollars for a single scientific cruise. Once a proposal for a scientific cruise has been funded, then the chief scientist has to obtain the equipment and organize all the logistics and scientific personnel for a cruise.

In the case of my first cruise, because I was the first student doing deep-sea marine geology at Oregon State, I had to design and obtain all of the laboratory and shipboard equipment (e.g. Fig. 2.9a). These logistics for a scientific cruise can be complicated, because the equipment has to be transported to the ship and sometimes this means sending it worldwide and being sure it arrives at the ship. If your equipment does not arrive at the ship in time, the chief scientist will lose tens of thousands of dollars of project and taxpayers dollars per day when the ship doesn't work.

The obtaining, organizing and supervising of the scientific staff is equally as difficult as taking care of equipment and logistics. On some of my scientific cruises I have supervised up to 40 scientists from 13 countries (e.g. Fig. 2.7b). Not only do you have to find and arrange for many scientists with different specialties, but you also have to obtain the electronic and mechanical technicians to keep the equipment running during the cruise. Putting the special equipment on the ship that you need for your cruise and keeping it running is a major challenge in the severe ocean environment. I have had so many electrical failures because of an incompetent chief engineer on a ship that finally no electrical equipment functioned anymore, including navigation, and the scientific cruise had to be aborted.

After all these obstacles to overcome and obtain accurate scientific information, it is extremely frustrating as a scientist to learn that a large percentage of USA people have become science deniers that do not believe in evolution, climate change or other scientific facts. This present change of attitude towards science has had severe consequences especially for natural science, when natural science studies are so critical to developing a sustainable Earth and avoiding global change problems. Because of this new cultural change towards science in the United States, budgets for natural science have been greatly reduced. In fact, in the 1994 contract with America, Congress proposed that the National Weather Service and the USGS should be elim-inated from the United States government. Citizens in the United States should remember that agencies like the Weather Service and the USGS provide important scientific data, such as detailed forecasts for hurricanes and the world's best information for earthquakes and other geologic hazards.

2.7   Sea Legs and Tsunami

November 1963 was the beginning of many oceanographic adventures that have continued for the rest of my scientific career. My first scientific cruise occurred during the stormy weather of the North Pacific Ocean. This cruise took place on the ship RV Acona, at that time the new oceanographic ship for Oregon State University. We left Newport, Oregon and headed north toward the Columbia River. As we traveled north we passed a small yacht with three men who waved as we traveled by. At midnight that same day we heard a mayday call on the radio for help from this boat. While trying to cross the bar, or entrance to the Newport harbor, the boat had capsized and all three lives were lost. This was a sobering lesson about the hostile ocean environment and the occasional dangerous scientific work that takes place on oceanographic cruises.

The boat capsized at Newport because a severe winter storm had arrived off the coast of Oregon and now our ship was in this storm. The RV Acona was known to be poorly designed and extremely unstable. For example during the rest of the week for my first cruise, much of the time the ship rolled 55° each way in the rough seas. Most ships would sink if they had this extreme roll, but the Acona, was like a cork bobbing in the sea. On one of its first cruises when crossing the Newport harbor entrance, where the waves reach shallower water and become extremely high, the Acona rolled flat on its side and lit all the emergency flares along the railing of the bridge. The ship bobbed up, but after this the captain was extremely careful when crossing the bar. As a result after my first stormy cruise, we had to wait two days to enter the harbor, because of the storm waves. As you will see this was terrible to wait after enduring the stormy sea for a week.

This first cruise was not a pleasant experience, because I was extremely seasick on this small 25 m (80 foot), unstable ship. Somewhere along the line my Norwegian Viking blood was lost and I did not have good sea legs. Not only that, but I was in a small laboratory with a retired U.S. Navy person who smoked big cigars. Also the early echo sounding equipment in this lab recorded the depth to the seafloor by electrically burning lines on the paper of the depth recorder. This resulted in large quantities of ozone filling the room along with the cigar smoke. In addition, my research work was to read temperatures on Nansen water bottles with a magnifying glass. If you have been on the sea, you know the advice is to look far in the distance and not close up at things, which helps creates seasickness.

The combination of rolling ship, cigar smoke, ozone and magnifying glass was a disaster for me and I became violently seasick, even though I was taking seasick pills. However, because of being sick, the pills never stayed in my stomach long, so I kept taking more. The result was I began hallucinating with the extreme seasickness and overdose of seasickness pills. When I finally reached land, I continued with land sickness with no equilibrium and more hallucinations for a week. After this first experience at sea it was amazing I still wanted to be an oceanographer.

On my second oceanographic cruise my sea legs were better, although the ship did worse. This was my first PhD cruise, and the maiden voyage of the new R.V. Yaquina research ship for Oregon State University. The ship was a great improvement, however all the equipment was being used for the first time. We lowered the piston corer and it triggered fine and penetrated several meters into the sea floor. Instead of pulling the corer out of the sea floor, the new winch on the ship failed to function. As a result, we were anchored for several days to the deep-sea bottom about 3,000 m (10,000 feet) below, while we tried to get the winch to function and pull the corer out of the sea floor. Our options then became cut and lose the 3,000 m of cable and piston corer, which were worth tens of thousands of dollars, or get everyone on the ship to try and turn the 20 ton winch drum. Just as we gathered everyone to try and turn the winch drum, it finally functioned. We saved our corer and cable, but that was the end of my first cruise to obtain PhD data. Fortunately, all my later cruises were succesful and I obtained excellent data that remains some of the best collected on a submarine fan.

While living in Oregon, I was introduced not only to the ship adventures of oceanography, but also to the natural global changes caused by earthquakes. First, while attending a night class in Portland Oregon, a small earthquake was centered under the old brick building where the class was taking place. The building shuck violently and all the lights went out, but there were no injuries or damage to the building. The second introduction to earthquakes was in April 1964 when I was attending a Geological Society of America meeting in Seattle. The world's second largest historical earthquake at that time took place in Alaska. This earthquake caused widespread destruction and damage in Alaska and tsunamis along western North America. I remember standing next to Dr. Frank Press and looking at the earthquake seismograph lines in Seattle. Dr. Press was a world authority on earthquakes and was later to become science advisor to President Clinton.

As soon as I returned from Seattle, I went to see the destruction of the tsunami on the Oregon coast. As well as the tsunami destruction, a number of lives were lost along the Oregon coast and particularly in Crescent City, California, just south of the Oregon border. Much of the low-lying region of Crescent City was completely destroyed. On the beaches and along the estuaries of Oregon, things were completely changed. Normally because of all the forests and logging in the Pacific Northwest, many driftwood logs accumulate on the back areas of the beaches. Like the people camping on the beach, the tsunamis had swept all of these logs out to sea along with the beach sand. The beach was barren of everything and there was evidence of the tsunami for kilometers along the estuaries.

The transoceanic tsunami from the 1964 Alaskan earthquake also resulted in an extensive tsunami warning system that was developed along the coast of the Pacific Northwest. If you now visit any of these beaches you'll see signs explaining the tsunami hazard and how to escape. There are warning sirens all along the coast to alert people to potential local and transoceanic global tsunamis after great earthquakes along the West Coast of North America and around the Pacific Ocean rim.

On another visit to Seattle a few years after the 1964 earthquake, I experienced the worst historical earthquake at the time in the Seattle area. I was in a motel room in the morning and suddenly the room started slowly swaying. I knew it was an earthquake, so I looked for something to get under to follow the duck and cover recommendation for earthquakes. However I could not find anything to get under, so I went and looked out the window. The room kept swaying for tens of seconds and I could look out the window and see trees also shaking and telephone poles moving back and forth with the electrical and phone lines whipping up and down.

This Seattle earthquake did not seem like a severe earthquake compared to the violent shaking that I encountered in my first earthquake experience in Portland. Consequently I assumed this was a minor earthquake, but was surprised after leaving the motel and walking down the street to see that some of the large plate glass windows were broken. I still did not think too much of it and went on with my work at University of Washington. After I finished, I drove back to Oregon and heard on the radio that several people had been killed and some buildings in the Seattle area had collapsed. So although the felt motion was not very severe, the length of the earthquake for tens of seconds did cause damage and a loss of life.

2.8   Around the World in 180 days

In 1966 the last year of my PhD program, I had the greatest introduction in my life to the global environment. I was fortunate to be hired as a professor of marine biology on the University of the Seven Seas, World Campus Afloat. I was a professor from the fall of 1966 to early 1967, while the ship circled the globe and visited the environments in many countries. When in port, I helped lead marine biology field trips with local scientists. These field trips significantly expanded my view of global environments, because I was able to visit worldwide locations of temperate and tropical areas.

Some of the interesting sites that we visited in Portugal were to observe effects of the 1755 Lisbon earthquake. We were shown damage and the high levels that tsunami waves reached in the city where an estimated 10,000 to 100,000 lost their lives and damage occurred throughout Portugal (https://en.wikipedia.org/wiki/1755_Lisbon_earthquake). This also was an early experience learning about the hazards, loss of life and damage related to earthquakes worldwide. Little did I realize that 40 years later during my retirement, my research would focus on defining great earthquake history in several locations around the world. The Lisbon earthquake caused the local tsunami and also deposited onshore tsunami sand layers on the Iberian coast. My wife and other Spanish scientists studied these onshore tsunami and seafloor earthquake sand layers to obtain a record of previous earthquakes, their frequency and potential hazards for the Iberian Peninsula.

After Lisbon Portugal, our next port stop was Barcelona Spain, where I had the good fortune of being guided on a field trip by Dr. Margalef. He was a world famous marine biologist at the Ciencias del Mar or Marine Science Institute in Barcelona and wrote the textbook that I used in my marine biology course at Oregon State University. On this field trip he explained the ecology of hake, also called merluza in Spanish, and expressed his concern at the diminishing numbers of fish related to the overfishing and increasing pollution of the Mediterranean Sea. The Mediterranean is similar to lakes, because this smaller confined part of the ocean is affected first by the human-caused changes. By 1966 the large populations and pollution from cities dumping human refuse, particularly untreated sewage, had resulted in severe pollution of the Barcelona area and other parts of the Mediterranean. Fortunately, the pollution problems have greatly improved, however the overfishing problem and loss of global fisheries still remains in the Mediterranean Sea and elsewhere (see details in Chap. 6 Sect. 6.8).

After Barcelona we traveled to Marseille France, but when we arrived in the harbor we were struck by the worst Mediterranean storm in centuries [7]. The wind was blowing more than 200 km/hour (120 miles an hour)) with sound screaming around the ship from these strong winds. While looking out the porthole window I could hear small boat engines. Then suddenly at the top of a wave, a small fishing boat would appear and then disappear way down in the trough of the wave and then appear at the top again. I never knew what happened to the fisherman and whether they survived this extreme storm.

Two days later when we arrived in Florence we could see huge damage on land from the worst flooding for hundreds of centuries in Florence [7]. This was not a typical storm and had I observed yet another extreme storm event possibly made worse by global warming. As I've explained before, you cannot relate one single storm, tornado, hurricane or flood to global warming. However when you see this worldwide pattern of intensified extreme storms over the past few decades, this is what the climate models have predicted as a result of global warming. There always have been severe storms in the Mediterranean Sea and elsewhere, but when the recent history of storms everywhere is compared to the past history of storms, there has been a change to more extreme storms such as the 2017 and 2018 hurricanes, typhoons and flooding in North America and Asia.

When the ship visited Egypt, we went to an oceanographic Institute. This was in late 1966 before the Aswan dam was completed. The Egyptian scientists were concerned about the environmental problems that could be caused by the daming of the Nile River. Once the Aswan dam was completed, the yearly floods of the Nile, which spread across the floodplain and the delta, would end. These yearly floods deposited river sediment, which enriched the farmland along the river and in the delta. Once the yearly deposits ceased, the farmland, which was the best in Egypt, would no longer be as productive for food.

The yearly floods of the Nile River also were the main source of nutrients for the Eastern Mediterranean Sea and sediment for building out the delta coastline. Once these yearly floods ended, the fisheries of the Mediterranean Sea would not be as productive and farmland would be reduced as the coastline eroded. Thus, the scientists thought that this lack of flooding would severely reduce the food production for Egypt, a country that already lacked sufficient food resources. This lack of food was evident as we traveled the streets of Cairo and saw many undernourished children.

Another problem, that concerned the Egyptian scientists, was if the river flooding ceased, the disease schistosomiasis would become much more prevalent. Apparently the river flooding hampered the cycle of snails, which were the main carrier of the disease. Actually, years later after the Aswan dam was fully operating, this disease decreased in Egypt (Abd-El Monsef 2015).

Unfortunately many of the other environmental problems that the Egyptian scientists were concerned about in 1966 have since resulted. Food production in the Eastern Mediterranean Sea has been reduced, the soil has become less fertile and contaminated with salts, and the delta coastline has eroded because of the human-caused global change of the Aswan Dam (Abd-El Monsef 2015).

There have been a number of global change problem caused by river dams. The dams on rivers of the west coast of America, especially the Columbia River in Oregon, have restricted the annual migration of salmon so that they cannot travel up river to spawn. Consequently, the wild salmon fisheries of Western North America have been greatly reduced. Engineers have tried to build fish ladders so that salmon can swim upstream and continue to spawn. However it was found that the water pouring over dams increased the nitrogen gas in the water below the dams and this resulted in the death of many salmon, in addition to fish dying trying to use the fish ladders (EPA technical report TS 09–70-208–016.1). Under the Obama administration, to help the use of fish ladders, the water outflow from Columbia River dams was increased to allow more salmon juveniles to migrate to the ocean (Seattle Post paper, May 26, 2011). The Trump administration has now introduced legislation to decrease the water flow. Why this administration wants to decrease salmon production is hard to understand, because better salmon runs help the fishing industry.

The lesson for global change that can be learned is that all of the natural costs and benefits need to be considered, as well as the value of irrigation water and electrical power that is gained by building a dam. Another lesson is the old adage that you cannot fool with Mother Nature. Once humans disturb natural systems, many changes are triggered and there are many actions and reactions which make it difficult to try and bring back a functioning natural system such as returning the salmon runs in the Columbia River system.

The next stop for the University of the Seven Seas was in Bombay, or now called Mumbai, India. The impact of major poverty was striking and an important lesson about global change and feeding the exponentially increasing global population. At the time India was facing famine, and many ships were anchored in the harbor waiting to unload food supplies to relieve the famine. Unfortunately, the government bureaucracy, corruption and lack of logistics prevented the ships from being unloaded for weeks. Meanwhile, people were starving in the streets, because many people lived in the streets without homes. We were told that around 300 bodies a day were picked up from the Mumbai streets in 1966. This experience brought home the lesson, that the problems of poverty and famine are not only related to human caused increased population, global warming, decreased farmland, and reduced water supplies, but also are caused by poor government policies.

The science of agriculture has greatly improved since 1966 with things like the green revolution, so that world food supplies have greatly improved since my 1966 visit to India. However, working against this is the loss of farmland because of extreme floods, the loss of groundwater for irrigation and global warming drought that expands desert areas. These problems are increasing, particularly in the sub-Sahara region of Africa where global warming and droughts have reduced food supplies and famine is increasing. Some of these drought problems could be relieved by irrigation from groundwater, but the use of groundwater for irrigation has not been sustainable globally. In fact with present rates of use, new studies indicate that the global groundwater for irrigation and drinking water supplies may be depleted within the next 50 years [40], NASA Jet Propulsion Laboratory News, June 16, 2015). This is yet another global change, among many, that has occurred during my lifetime.

After India, we went on to Ceylon (since called Sri Lanka), where I had my first introduction to one of the Earth's most astounding environments, coral reefs. We visited the coral reefs and went skin diving amid the great numbers of brightly colored fish and corals. All of the fish were completely unafraid and were not bothered by the presence of humans. You could swim among the fish and were accepted as part of their environment. They would come up and touch you and did not swim away frightened when you came into their coral reef territory. At that time in 1966, few tourists visited these remote areas of coral reefs. Later on the ships stop in Hawaii, there was a striking difference while I was skin diving in coral reefs. There were many people snorkeling and spearing fish along the coral reef. Even in 1966, the Hawiian fish were very timid compared to the coral reefs of Sri Lanka.

Unfortunately, these trends for the human caused destruction of coral reefs have continued during my lifetime because of the global population increase that has resulted in overfishing, pollution and global warming. A quarter of all marine fish species reside in coral reefs and 500 million people depend on these "underwater rainforests" for their livelihood [41]. However as populations have increased, so has the over fishing. In fact in some places, the fishing methods are so extreme that dynamite is exploded along the reefs to stun the fish and then gather up huge amounts of the stunned fish.

Another cause for the loss of coral reefs is the increase of CO_2 that results in ocean acidification and disolving coral, and global warming which increases the sea temperature and causes coral bleaching (e.g. [13]). If you are not familiar with the concepts of global warming, ocean acidification, and coral bleaching, see the detailed descriptions in Part II Chap. 5, Sects. 5.2 to 5.4 and Chap. 6, Sect. 6.8. When returning to Malaysia in 2016 after 50 years, I observed examples of bleaching and many other ecological changes to reefs and mangroves. Recent studies suggest that the combination of over-fishing, pollution, global warming and ocean acidification will result in the destruction of many coral reefs within the next 40 years [41]. Fortunately, other recent coral reef assessments found that 45 percent of the world's reefs are healthy, providing hope that a few reefs may be able to endure the changes expected from global warming.

Another new and fascinating environment of mangroves was explored in 1966 during our port stop in Malaysia (https://www.vizcaino.cat/en/mangro ves/). We visited a mangrove area that had grown into the sea for 50 km since Marco Polo explored this area around 1292 A.D. This amount of growth into the sea is shown by the fact that the port that Marco Polo visited was 50 km inland from the edge of the mangroves that I visited in 1966. The mangroves had increased the land surface because they trap sedi-ment to extend the land into the sea. Mangroves are inhabited by a wide variety of species such as bright colored crabs that crawl through the mud in the mangrove roots. Mangroves are important for sequestering carbon from global warming and providing a buffer from storm waves in low lying coastal areas that are threatened by rising sea levels [6].

Unfortunately, similar to coral reefs, the mangrove forest areas have decreased 20% between 1980 and 2005, although fortunately this forest loss has since slowed [46]. The main destruction results from humans rapidly cutting mangroves for wood and expanded land for crops. The loss of mangroves results in land area becoming eroded away by the sea and reduced protection from extreme storm waves. Consequently areas like I visited in 1966 have now been receding back many kilometers compared to the 50 km of growth after Marco Polo visited Malaysia.

When we visited Japan, we observed the dense population along the coast, which requires increased land area for building, Consequently, Japan has reclaimed large areas of the sea along Tokyo Bay by depositing artificial landfill. Unfortunately, this artificial landfill is unstable during great earth-quakes that are common along the Japanese coast. The most recent example is the 2011 earthquake to the north of Tokyo, where thousands of lives were lost. An even greater loss of life occurred in the 1923 earthquake of Tokyo. Consequently, this artificial landfill area that now is covered with buildings in Tokyo, is at extreme risk for a future earthquake. The Japanese society and government are aware of these earthquake hazards and have developed the most sophisticated infrastructure to try and cope with the extreme earthquake hazards they face. For example, when an earthquake takes place, within a few seconds all high-speed trains are stopped so that train wrecks do not occur. Also all bridges have been refitted so that even with the strongest earthquakes the bridges will not collapse and can be used immediately after the earthquake.

Even with all the Japanese earthquake planning, their geological research has not always been utilized. The 2011 earthquake and extremely destructive tsunami inundated and destroyed much of the Fukushima power plant and released radioactive pollution. The geological studies, prior to 2011, showed that about every 800 years, tsunamis swept past the area of the Fukushima nuclear plant and traveled several kilometers inland [8]. However the government did not utilize this information and built the nuclear plant in a hazardous location. Again this is a global lesson that has not been learned in many areas, because although geological information exists, there has not been political will to spend funds for infrastructure that will prevent human deaths and property destruction.

There also is a global problem in that the newest information about tsunamis has not reached remote areas. The worst example of this was the 2004 Sumatra earthquake when an estimated 227,898 people in 14 countries lost their lives, mainly because of tsunamis (https://en.wikipedia.org/wiki/2004_Indian_Ocean_earthquake_and_tsunami). The information to rapidly retreat to higher ground to avoid tsunamis was not generally known in Sumatra, even though a history of earthquakes was known. In fact there was only one coastal island in Sumatra where the population retreated to higher ground, because this knowledge was passed down in the folklore. A sad but ironic story about Sumatra was that of Dr. Kerry Seih, a California Institute of Technology professor. He had studied earthquakes in Sumatra for years and wanted to educate the population about tsunami hazards. He personally printed up a number of warning signs to bring with him and post in coastal areas of Sumatra in January 2005, but tragically the earthquake took place a month before his visit.

References

1. Adams J (1990) Paleoseismicity of the Cascadia subduction zone—Evidence from turbidites off the Oregon-Washington margin. Tectonics 9:569–583
2. Alvarez LW, Alvarez W, Asaro F, Michel HV (1980) Extraterrestrial cause for the cretaceous-tertiary extinction. Science 208:1095–1108

3. Byron Earl R, Charles R Goldman (1984) Recent sedimentation and the fertility of Lake Taho, A report to the Interagency Tahoe Monitoring Program, Tahoe Research Group, Inst. Ecology, Univ. California, Davis, p 26

4. Byron ER, Goldman CR (1986) Long-term changes in the atmospheric deposition of acid and nutrients at Lake Tahoe, CA-NV, EOS. Trans Amer Geophysical Union 67:986

5. Ceballos G, Ehrlich PR, Rodolfo D (2017) Biological annihilation via the ongoing sixth mass extinction signaled by vertebrate population losses and declines. Proc Nat Acac Sci Am 114:E6089–E6096. https://doi.org/10.1073/pnas.1704949114

6. Cleverley M (2015) Amazing mangroves. Pacific Ecol 23:28–31

7. DeZolt S, Lionello P, Nuhu A, Tomasin A (2006) The disastrous storm of 4 November 1966 on Italy. Nat Hazards Earth Syst Sci 6:861–879. www.nat-haz ards-earth-syst-sci.net/6/861/2006/

8. Goldfinger C, Ikeda Y, Yeats RS, Ren J (2013) Superquakes and supercycles. Seismol Res Lett 84:24–32. https://doi.org/10.1785/0220110135

9. Goldman CR (1973) Will Baikal and Tahoe be saved? Cry California. Winter 1973–74:19–25

10. Goldman CR, Richards RC (1987) The urbanization of the Lake Tahoe basin: a microcosm for the study of environmental change with continuing development. In: Bradley D (ed.) Proc., State of the Sierra Symp. 1985–86, Pacific Publications, San Francisco, pp 42–62

11. Gould LM (1931) Cold: the Record of an Antarctic Sledge Journey, Brewer, Warren & Putnam

12. Hansen J, Makiko S, Reto R, Gavin A Schmidt, Lob K, Persin A.(2018) Global Temperature in 2017. http://www.columbia.edu/~jeh1/mailings/2019/20190206_Temperature2018.pdf

13. Hughes T, Kerry J, Baird A, Connolly S, Dietzel A, Eakin C, Heron S, Hoey A, Hoogenboom M, Liu G, McWilliam M, Pears R, Pratchett M, Skirving W, Stella J, Torda G (2018) Global warming transforms coral reef assemblages. Nature 556:492–496

14. Kolpin DW, Furlong ET, Meyer MT, Thurman EM, Zaugg SD, Barber LB, Buxtpn HT (2002) Pharmaceuticals, hormones, and other organic wastewater contaminants in U.S. streams, 1999–2000: a national reconnaissance. Environ Sci Technol 36:1202–1211. https://doi.org/10.1021/es011055j

15. Kunkel KE. Plus 23 co-authors, 2013, Monitoring and understanding changes in extreme storms: state of knowledge, Bull. Am. Meteorol. Soc., 499–514. DOI:https://doi.org/10.1175/BAMS-D-12-00066.1

16. Malthus TR (1798) An essay on the principle of population, anonomous essay later identified with Malthus as the author

17. Meadows DH, Meadows DL, Randers J, Behrens III, William W (1972) The limits to growth; a report for the club of Rome's project on the predicament of mankind. Universe Books, New York. ISBN 0876631650

18. Meadows DH, Randers J, Meadows DL (2004) The limits to growth: the 30-year update. White River Junction VT, Chelsea Green Publishing Co. ISBN 1931498512
19. Miller DJ (1960) Giant waves. In: Lituya Bay, Alaska, U.S. Geological Survey Professional Paper 354- C, DOI, https://doi.org/10.3133/pp354C
20. Morey AE, Goldfinger C, Briles CE, Gavin DG, Colombaroli D, Kusler JE (2013) Are great Cascadia earthquakes recorded in the sedimentary records from small forearc lakes? Nat Hazards Earth Syst Sci 13:2441–2463. https://doi.org/10.5194/nhess-13-2441-2013
21. Moernaut J, plus 7 other authors (2007) Giant earthquakes in South-Central Chile revealed by Holocene mass-wasting events in Lake Puyehue. Sedimentary Geol 195:239–256. DOI: https://doi.org/10.1016/j.sedgeo.2006.08.005
22. Nelson CH (1967) Sediments of Crater Lake, Oregon. Geol Soc Am Bull 78:833–848
23. Nelson CH (1976) Late Pleistocene and Holocene depositional trends, processes and history of Astoria deep-sea fan, northeast Pacific. Mar. Geol 20:129–173
24. Nelson CH, Bacon CR, Robinson SW, Adam DP, Bradbury JP, Barber JH Jr, Schwartz D, Vagenas G (1994) The volcanic, sedimentologic and paleolimnologic history of the Crater Lake caldera floor, Oregon: evidence for small caldera evolution. Geol Soc Am Bull 106:684–704
25. Nelson CH, Baraza J, Maldonado A, Rodero J, Escutia C, Barber Jr JH (1999) Influence of the Atlantic inflow and Mediterranean outflow currents on Late Quaternary sedimentary facies of the Gulf of Cadiz continental margin, In: Mandonado A, Nelson CH (ed.) Marine Geology of the Gulf of Cadiz. Mar. Geol. Special Issue, v 155, pp 99–129
26. Nelson CH, Carlson PR, Bacon CR (1988) The Mt. Mazama climactic eruption and resulting convulsive sedimentation on the continent, ocean basin, and Crater Lake caldera floor. In: Clifton HE (ed) Sedimentologic Consequences of Convulsive Geologic Events, Geological Society of America Special Paper, 229, pp 37–56
27. Nelson CH, Carlson PR, Byrne JV, Alpha TR (1970) Physiography of the Astoria Canyon-Fan system. Mar Geol 8:259–291
28. Nelson CH, Escutia C, Karabanov EB, Colman SM (2000) Tectonic and sediment supply control of deep rift lake turbidite systems. Lake Baikal, Russia, Reply, Geology 28:190–191
29. Nelson CH, Goldfinger C, Johnson JE, Dunhill G (2000) Variation of modern turbidite systems along the subduction zone margin of cascadia basin and implications for turbidite reservoir beds. In: Weimer PW, Nelson CH et al. (eds.) Deep-water Reservoirs of the World, Gulf Coast Section Society of Economic Paleontologists and Mineralogists Foundation 20th Annual Research Conference, CD ROM, p 714-738

30. Nelson CH, Goldfinger C, Gutierrez-Pastor J (2012) Great earthquakes along the western United States continental margin: implications for hazards, stratigraphy and turbidite lithology. In: Pantosti D, Gràcia E, Lamarche G, Nelson CH editors, Special Issue on Marine and Lake Paleoseismology, Nat. Hazards Earth Syst. Sci., 12:3191–3208

31. Nelson CH, Kulm LD, Carlson PR, Duncan JR (1968) Mazama ash in the northeastern Pacific. Science 161:47–49

32. Nelson CH, Meyer AW, Thor D, Larsen M (1986) Crater Lake, Oregon: a restricted basin with base-of-slope aprons of nonchannelized turbidites. Geology 14:238–241

33. Nelson CH, Nilsen TH (1984) Modern and ancient deep-sea fan sedimentation. Society of Economic Geology and Paleontology Short Course 14, Tulsa, OK, p 403

34. Netflex (2020b) Kiss the Ground. https://www.netflix.com/es-en/title/813 21999

35. Polonia A, Nelson CH, Romano S, Vaiani SC, Colizza E, Gasparotto G, Gasperini L (2017) A depositional model for seismo-turbidites in confined basins based on Ionian Sea deposits. Marine Geol 384:177–198. ISSN 0025–3227, https://doi.org/10.1016/j.margeo.2016.05.010

36. Satake K, Shimazaki K, Tsuji Y, Ueda K (1996) Time and size of a giant earthquake in Cascadia inferred from Japanese tsunami records of January, 1700. Nature 379:246–249

37. Slotton DG, Goldman CR, Axler RP, Reuter JE (1987) Mercury accumulation in a new reservoir system. In: Lindberg SE, Hutchinson TC (eds.) Heavy metals in the environment. Page Bros., Norwich, Great Britian, pp 63–65

38. United Nations (2019a) Intergovernmental Science-Policy Platform on Biodiversity and Ecosystem Services (IPBES) Global Assessment Summary for Policymakers. www.ipbes.net/news/ipbes-global-assessment-summary-policy makers-pdf

39. United Nations (2019b) World population prospects, the 2008 revision

40. Wada Y, van Beek LPH, van Kempen CM, Reckman JWTM, Vasak S, Bierkens MFP (2010) Global depletion of groundwater resources. Geophys Res Lett 37:L20402. https://doi.org/10.1029/2010GL044571

41. World Watch Institute (2018). http://www.worldwatch.org/node/5960

42. Zdwiczano CM, Zielinski GA, Germani MS (1999) Mount Mazama eruption—Calendrical age verified and atmospheric impact assessed. Geology 27:621–624

43. Wing OEJ, Bates PD, Smith AM, Sampson CC, Johnson KA, Fargione J, Morefield P (2018) Estimates of present and future flood risk in the conterminous United States. Environ Res Lett 13(7):034023. https://doi.org/10.1088/1748-9326/aaac65

44. Goldfinger C, Nelson CH, Morey AE, Gutierrez-Pastor J, Johnson JE, Karabanov E, Chaytor J, Dunhill G, Ericsson A (2012) Rupture lengths and temporal history of Cascadia great earthquakes based on turbidite stratigraphy,

USGS Professional Paper 1661-F. In: Kayen R (ed) Earthquake hazards of the Pacific Northwest Coastal and Marine regions, p 192

45. Wilson EO (2002) The future of life. Knopf Doubleday Publishing, New York City, p 220

46. FAO Newsroom (2008) Mangrove loss http://www.fao.org/newsroom/en/news/2008/1000776/index.html

47. Wentz DA, Brigham ME, Chasar LC, Lutz MA, Krabbenhoft, DP (2014) Mercury in the Nation's streams—Levels, trends, and implications: U.S. Geological Survey Circular 1395, p 90. https://doi.org/10.3133/cir1395

48. Tibbetts JH. (2018) Air-Gun Blasts Harm Marine Life across the Food Web. BioScience, 68:1024. https://doi.org/10.1093/biosci/biy123

3

Alaskan Lessons

3.1 Earthquake and Ignored Geologic Warnings

In the summer of 1966, I was hired for a permanent job with the new Marine Geology Branch in the Geological Division of the United States Geological Survey (USGS). This Division is the research division of the USGS and has the main missions of resource research, environmental assessment and determining geologic hazards. The USGS hired me to help with their first research cruise to Alaska because President Lyndon Johnson decided that the way to help the US balance of trade payments was to find gold. Thus the USGS developed a marine geology program to find gold in offshore Alaska and my first USGS task was to explore for gold along Alaskan coasts (Fig. 3.1).

On my way for Arctic gold studies, I spent my first night in Anchorage, Alaska (Fig. 3.2). There I had an immediate encounter with natural global change, because it was only two years since the earthquake of 1964 had destroyed this city and killed people there and in tsunamis along the coast of Oregon and California. Photos taken just after the earthquake showed large areas of Anchorage where the tops of buildings had dropped down to the street level and destroyed housing areas because of landslides. When I arrived, the downtown had been leveled so it was just open spaces with no buildings, which resulted in a strange view out of my hotel window. Not only was there great local destruction and loss of life in Alaska, but also there were global implications from the tsunami that traveled across the ocean to western

© The Author(s), under exclusive license to Springer Nature
Switzerland AG 2021
C. H. Nelson, *Witness To A Changing Earth*,
https://doi.org/10.1007/978-3-030-71811-4_3

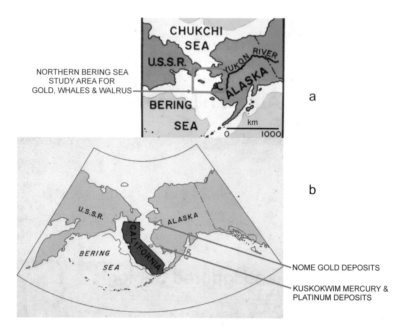

Fig. 3.1 a Map of Alaska showing the area (pink box) of my 1960s, 1970s and 1980s studies in the northern Bering Sea. Note in the southeast corner of the pink box is the location of the Yukon River Delta. The light blue offshore area is the continental shelf, which is less than about 120 m (400 feet) of water depth, where my studies took place. **b** Map showing the location of the Kuskokwim area of mercury and platinum deposits and the Nome Alaskan offshore area where my USGS study discovered the world's largest marine gold deposits. The area of California is superimposed to emphasize the enormous size of the Alaskan USA offshore area (2/3 of total USA offshore area). These maps show the USSR at the time of my studies during the cold war. After that USSR was divided into Russia and the Commonwealth of Independent States (CIS) as shown in Fig. 3.2. (https://en.wikipedia.org/wiki/Commonwealth_of_Independent_States). Figure a is from Hans Nelson and b is from Mike Marlow U. S Geological Survey, Menlo Park, CA

North American coasts. There also was damage in the Gulf of Mexico where earthquake waves traveling globally caused the Gulf water to slosh back and forth, which resulted in boats breaking loose from their moorings in harbors.

As mentioned before how governments do not heed geologist's hazard studies, the USGS geologists had warned decades before that Anchorage was extremely vulnerable to earthquake damage and of course Anchorage was in a major earthquake prone area. The USGS geologists had shown that consolidated rock layers overlay an unstable bootlegger cove clay layer along which landslides could take place. With the tremendous shaking of the 9.2 magnitude 1964 Alaskan earthquake, the second strongest historical earthquake at the time, landslides took place just as predicted. In the following years as

Fig. 3.2 Map of Alaska showing offshore locations in Bering Sea. The blue labels are names of bodies of water and the black labels, except for Bering Strait and the International Date Line are features on land. Norton Basin is the deep subsurface basin underlying Norton Sound. Cape Prince of Wales is the western point of Alaska on Bering Strait and Chukotka is the eastern point of Russia on Bering Strait. Chirikov Basin is the area between St. Lawrence Island and Bering Strait. Kotzebue village is located on the north side of Kotzebue Sound and Deering village is located on the south side of Kotzebue Sound. Map source is GraphicMap.com

I traveled through Anchorage, I observed that many of the areas destroyed by the earthquake in the town center, year by year were gradually rebuilt. This was not true of the destroyed housing areas which now remain as Earthquake Park in Anchorage where you can observe the landslide area with good explanation signs.

Previously I mentioned the geological warnings about tsunami hazards in Japan that were not considered when the Fukushima nuclear power plants were constructed. I just mentioned a similar case of earthquake hazards in Anchorage, where geologist's warnings also were not heeded. Unfortunately, this is a global change problem, because as the population increases, more people live in known hazardous areas, especially along rivers with flooding, in coastal regions with hurricanes and around the Pacific Ocean rim with earthquakes.

Another tragic event that I remember was the Vjont dam disaster of 1963 in Italy, where geologist's warnings also were ignored. Years later on a field trip to Italy, I visited the dam site, and observed the geological setting that had been warned about prior to the construction of this 262 m (860 feet) high dam in a deep narrow mountain canyon (https://en.wikipedia.org/wiki/ Vajont_Dam). High mountains rose above the location where the dam was constructed and geologists noted that the south mountainside was unstable. The dam was constructed anyway and as the reservoir began filling a landslide took place in 1960. Nevertheless, they resumed filling the reservoir and then a 2,000 m crack appeared in the mountainside above the dam, which signaled that the mountain could have a huge landslide. Unfortunately, they continued filling the reservoir, and as predicted, when the reservoir was almost full, the mountainside slid into the reservoir and forced all of the water to overtop the dam in a 252 m (820 feet) high wave. The water roared as a flash flood through the mountain valley and completely destroyed several towns below the dam. More than 1,900 people were killed in this flash flood that could have been prevented if geologist's warnings had been heeded.

3.2   Looking for Offshore Gold

My first work for USGS in Alaska was to look for offshore gold deposits in the northern Bering Sea and southern Chukchi Sea during the summer of 1966 (Fig. 4.1). I describe this research as an example of how the majority of geologists work in exploration for mineral and petroleum resources. These resources have powered the industrial revolution, but also have had a major impact for global change such as the input of CO_2 from burning fossil fuels, which is causing global warming. President Johnson increased funding for the USGS in the hope that the USGS would find gold to solve the USA global balance of payments problem. Even though I did find the world's largest offshore gold deposit, this was of little help because of the huge increases in petroleum imports that made the USA balance of payments soar so that finding gold would never solve the problem.

To begin our USGS gold project, we mainly sampled beaches around Kotzebue Sound, but did not locate any gold resources (Fig. 3.2). The main thing that I learned in 1966 was about the complex logistics problems that need to be overcome to work in the Arctic seas. This resulted in some crazy adventures that are described in the following section on 1960s Alaskan adventures.

In 1967 my assignment was to continue looking for offshore gold in the Nome Alaska area, where I spent most of the next 15 summer field seasons Figs. 3.1b and 3.2). Nome was the area that had the largest gold rush in Alaska at the turn of the century. The original discovery of gold was on the modern beach of the Nome Alaska shoreline. The first miners on the modern beach staked their claim, which was the area that they could swing their shovel and drive other miners away. No mother load of gold was found near Nome Alaska, because glaciers had scoured out the Snake River valley and pushed glacial deposits with gold to the shoreline and offshore [1]. Thus, when the waves eroded these glacial deposits, gold was concentrated at the base of the beach sand on top of the glacial clay, which the miners called false bedrock.

Soon after the initial modern beach gold discoveries, it was found that older sea levels of the past 5 million years had developed ancient beaches inland from the modern beach. These were called second beach and third beach and eventually large gold-mining dredges mined along these ancient beach ridges. An interesting side note is that in the Pliocene time from 2 to 5 million years ago, the climate was much warmer and the CO_2 amounts in the atmosphere were equal to what humans and the industrial revolution have now added to the atmosphere during the past 100 years [2]. These ancient beaches in Nome Alaska and globally are up to 20 m above the present sea level. They serve as a warning that global warming by the end of the twenty-first century can result in sea levels up to several feet higher than at present and affect about 20% of the global population.

Because we knew this history of ancient onshore beaches, the USGS and the US Bureau of Mines developed a project for the summer of 1967 to drill ancient offshore beach deposits the USGS had mapped out to the three-mile limit of state lands off Nome, Alaska (Fig. 3.1b and 3.2). Our government agencies also partnered with the Shell Oil Company mineral resources groups. In the early 1960s, Shell Oil vice president had the idea that there might be gold offshore and Shell Oil bought up all of the state offshore leases near Nome Alaska. As a result, the government groups and Shell Oil obtained a ship, the Virginia City, and outfitted it with a drill rig. For three months we drilled in the Nome Alaska offshore areas. Our methods were to drill 2 m (6 feet) and then obtain the unconsolidated sediment of the drill cuttings to sample for gold. This sample was then concentrated with the classical method of gold panning and then the panned gold was weighed. One of my jobs was to make a drill hole column with the gold values for each 2 m section down the drill hole.

I soon noticed that the highest values of gold were always in the first 2 m sample of the drilling. Because of this, I began taking samples around the ship while we were anchored for a day or two at each drill site. I walked all around the edge of the ship with a small dredging sampler on a rope and took near- surface samples of the seafloor. My idea turned out to be the most important finding of our summer's work. At first, the results were puzzling because many samples had little or no gold and others would have higher values with particles of gold about 1 mm in diameter. When analyzing this with others at the USGS, we realized that the small samples that I took were not large enough to represent the average value of the gold content.

A much larger sediment sample needed to be taken to accurately determine the average gold content, because of the size of the gold particles. However, I realized that there was a statistical way to increase my sediment sample size without taking larger samples. What I did was to make a moving average or combine the gold values of three sediment samples in a row and keep taking three samples at a time throughout the entire Nome offshore area. By doing this, I could make a more accurate analysis for the average gold values of the offshore area and thus map the high-value and low value areas. In the end when I wrote a USGS Professional Paper, I estimated from my 1967 research that there was a large area off Nome which contained about 1,000,000 oz of gold [1].

At that time in the late 1960s, when gold had a value of $35 an ounce, the value of this deposit was $35 million dollars. Years later and recently when the value of gold reached nearly $2000 per ounce, the value of my discovery became worth close to 2 billion dollars. My discovery remains the largest global resource of offshore gold. Consequently, in my first USGS project, I made a discovery that paid for my entire career. However, this was not of any financial benefit to the USGS, or me because Shell Oil owned the offshore leases.

The later history of my discovery was interesting because Shell Oil eliminated their mineral resource group and sold their Nome offshore leases to ASARCO mining company (American Smelting and Refining Company). A few years after my discovery, ASARCO took a barge with a large crane and bucket that could sample 1 m^3 (35 cubic feet) of seafloor sediment and confirmed that my statistical estimates of the gold value were accurate. ASARCO did not mine the deposit because they were only interested in developing mines with the value of one billion dollars or more. With the gold prices at that time, my discovery of the offshore gold deposit did not have enough value for ASARCO to mine.

Eventually during the end of apartheid in South Africa, an African gold mining company bought the Nome Alaska offshore leases from ASARCO. The African company bought the world's largest tin mining dredge in Malaysia and brought it to Nome Alaska. This dredge could mine down to 300 feet, but the African company after the first year modified the dredge to mainly mine the near-surface sediment, which I had showed to contain the highest content of gold. At that time after prices of gold had raised somewhat, the African company hoped to mine $25 million worth of gold a year. Because they only averaged about $17 million a year, after five years they abandoned the project. There are still companies investigating whether it is possible to mine the remaining gold off Nome Alaska.

The other interesting aspect of my 1967 gold discovery was to watch year-by-year how gold fever resulted in very crazy activities of many businessmen that hoped to get rich by mining offshore gold near Nome. This continues to this day, because there is even a Discovery Channel program called Gold Fever that shows miners with small ships and local people diving off Nome Alaska with small suction-dredge equipment to mine gold (e.g. gold fever T9, Episode 11). Shortly after my discovery, with little knowledge, many strange schemes were funded such as a monstrous tractor that would drive near shore to mine gold, but this was a failure.

The worst mining venture example was a Texas restaurant man who arrived in my USGS office one day at 11:45. He said that by 12:00 he had to decide about buying a mining ship in Seattle. It was the first time he had even considered the geology of the area and he only was going to spend 15 minutes on discussing this with me. I showed him that the area where he wanted to mine in Norton Sound was covered by Yukon mud and contained little or no gold (Fig. 3.2). He proceeded anyway and ended up with three ships for his mining venture in the wrong area. As I pointed out earlier, sudden storms are common in Alaska and one night all three of his ships sunk in a storm.

This is an important lesson about working in Arctic Alaska, because for 5000 miles of coastline north of the Aleutian Islands there are no harbors to shelter large ships (Fig. 3.2). This makes logistics extremely difficult when working in the Arctic seas. You cannot seek the shelter of a harbor, and to load equipment on and off a ship you have to put the equipment on a small boat and take it out to a ship anchored offshore in the open ocean. As you will see later in my descriptions of Alaskan adventure, there are many problems because of bad weather when you cannot take equipment or people to and from a ship in a harbor.

3.3 Hard and Humorous Logistics Lessons

My first adventure with Alaskan logistics was to try and find the Office of Naval Research (ONR) boat that we were to use for the summer research in 1966. This small 30-foot boat had been sent to Seattle, Washington from Barrow, Alaska to refit the engine (Fig. 3.2). The boat then was put on a barge that was pulled by tugboat to Alaska. However, no one was sure where the barge was located along the Alaskan coast. After several days of investigation, we found out that the barge was in Port Clarence Sound about 70 km (45 miles) northwest of Nome (Fig. 3.2). Unfortunately, a crane operator had to be brought from Anchorage about 1000 miles away by plane to get to the barge and unload our boat.

As is common for Alaskan logistics, before the flight of the crane operator, a part for his plane engine had to be obtained from the United States or lower 48 as the Alaskans say. After more than a week's time, the crane operator got to the barge and an ONR plane flew the Alaskan Native boat captain Frank and me to Port Clarence where we arrived at midnight. By this time, strong winds had begun and even in the 50 km (30 miles) wide Port Clarence Sound, significant waves were developing. We took a small outboard motor boat crashing through the waves out to the barge in the center of Port Clarence Sound. They lowered our small research boat from the barge and filled it with diesel fuel. The pilot and plane flew back to Nome, and captain Frank and I began chugging through the waves to reach the quiet harbor of Teller Alaska (Fig. 3.2). However, this was just the beginning of a long weekend adventure trying to reach the Teller, Alaska harbor.

As we chugged through the storm about half way across the Port Clearance Sound the engine died. We kept trying to start the engine until the large 200-lb batteries below deck where dead. We had a generator onboard and would lift these 200-lb batteries through the small hole in the deck and recharge them on deck. For hours and then days we kept up this routine of trying to start the engine and running down the batteries and then charging them again. During this time, we had no food or water. While trying to start the engine, we had additional problems because when they put in the new engine in the boat they did not label any of the controls. So captain Frank kept trying to figure out what the different controls were.

Captain Frank also discovered that the fuel they had given us contained water. This water then got into the injectors of the diesel engine and of course with water in the fuel injection system of the engine, it would not start. After bleeding all the water from the injectors, the engine still would not start. Finally, captain Frank figured out that one of the control buttons was an emergency stop button that had been pushed in. After about four days with no food and water, we managed to get the engine started again and began traveling across the rest of Port Clarence Sound to the Teller village harbor. At about the same time, the ONR plane pilot, after spending the weekend drinking in Nome, arrived back and began traveling in a small outboard motor boat to the research boat that we were on. You can understand that we were less than happy about the pilot's lack of concern for us, since he was supposed to wait and see that we got to shore after our boat was taken off the barge.

We spent about a week in Teller to outfit the boat with all of our equipment. Eventually, we had all of our equipment installed, which included echo sounders to determine the water depth, and high-resolution seismic systems (Fig. 2.5). We also had sediment sampling equipment and radar to help with navigation.

We left Teller harbor to head up to the Bering Strait, then into the Chukchi Sea and on to the town of Kotzebue at the eastern end of Kotzebue Sound (Fig. 3.2). It was a beautiful sunny day, however there were strong winds blowing about 50 knots (90 km or 56 miles per hour) to the south through Bering Strait. The Bering Sea became extremely rough as we came closer to the strait, because the oceanic Alaskan Coastal Current traveled north through Bering Strait at speeds up to 5 knots (10 km or 6 miles per hour) against the strong wind. Essentially it was like being in a surf zone on a stormy day, even though it was a bright sunny day, except for the wind. Our small research boat was tossed about like a cork and our equipment was all thrown from the shelves and smashed on the floor. Even one of the wheelhouse windows was broken so that seawater entered our small cabin area adding to the destruction of equipment.

To make matters worse we still had water in our fuel so that if the engine stopped, we would quickly be pushed to the shoreline and have the boat smashed to pieces on the rocks. Thus, captain Frank said I had to steer the boat while he was below deck continually bleeding water from the fuel filters to keep the engine from stopping. I have driven boats, but never this big and because of the storm I kept steering the boat close to the shore. Captain Frank said stay away from the shore so we do not hit the rocks. After several hours of fighting our way through the huge waves in the Bering Strait, we eventually reached the calmer water of the Chukchi Sea (Fig. 3.2). To show how high the waves were smashing over the boat in Bering Strait, a week or so later we noticed that the light at the top of our 10 m (30-foot) mast did not work. When we opened it, we saw it was full of seawater from the waves smashing over the boat.

We continued on into the night, which had daylight because of the midnight sun and we were north of the Arctic Circle (Fig. 3.2). We crossed Kotzebue Sound and reached the eastern end of Kotzebue village about 2 AM in the morning (Fig. 3.2). We woke up a husky dog team at this far end of the village. This team began howling which woke up the next team to the west which began howling and woke up the next team further west and this chain of mournful husky howling traveled eventually 3 km (2 miles) west-ward to the other end of the village. Then the whole process of husky dog team howling traveled from the west back to the east end of the village. This probably went on for about an hour and was a sound I will never forget. We then slept far into the morning like the Alaskan Natives did because with the midnight sun the whole village including small children would stay up late into the night. We now settled into our regime on our small research boat with captain Frank as the cook and me as the crew. Our food consisted of things like muktuk, which is rubberlike seal blubber, and uncooked frozen reindeer steaks.

After we got up and had our meal, we found that the boat could not travel away from the deeper channel that ran parallel to the shoreline. Kotzebue Sound, which is about 50 km (30 miles) across, is extremely shallow away from the few channels. For example, miles from the shore where no land is visible you can see seabirds that appear to be walking on the water, when actually they are on the seafloor which is only covered by a few centimeters or inches of water. To cross the Sound, boats have to follow a narrow channel to get to the village of Kotzebue. We were lucky the night before, because the strong winds that caused a storm surge had raised the water level across the shallow Kotzebue Sound and we could cross the sound without using the channel. However, the next day the water level had dropped back to normal and now our boat could not leave the deeper channel we were in by the shore-line. Captain Frank kept trying to find a way out from the deeper channel for several days, but with no luck. Because he was a Point Barrow Alaska Native, the Kotzebue Natives made fun of his lack of knowledge about the Kotzebue Sound channels. As a result, he got into a bar fight and was beaten up, which added to our problems.

After several days, when I had been staying with our bush pilot Bob Baker's family, I got up in the morning to find that captain Frank and the boat had disappeared. This disappearance lasted several days and then one morning I found the boat in the Kotzebue harbor. I never found out what captain Frank did to get back across Kotzebue Sound and into the channel to find his way into the Kotzebue harbor. I suspected to leave the channel where the boat was trapped, he put the boat in reverse and let the boat propeller spin and dig his way out of the channel near the shoreline and across the sound to find the main channel.

A new adventure was to begin once we had the boat prepared to begin research in the Kotzebue Sound area (Fig. 3.2). Because most of our equip-ment had been destroyed in the Bering Strait storm, we were left with one remaining type of geophysical equipment. This equipment consisted of two magnetometers, which determine magnetic field strength. Our idea was that seafloor deposits with valuable heavy metals like gold could be detected because other minerals with magnetic properties like iron minerals would be associated with them. Our system was to put the two magnetometers in different small boats in a catamaran arrangement so that they would determine the magnetic fields and identify heavy metal deposits under the seafloor.

Finally, we left Kotzebue to travel across to the south side of Kotzebue Sound to the small Alaska Native village of Deering, Alaska (Fig. 3.2). This was after several lost weeks of looking for our research boat, surviving the Bering Strait storm, and preparing our boat. We started our crossing into the beautiful red midnight sun on a perfectly calm glassy sea. Shortly after we began, as is common in Alaska, a sudden storm appeared and then storm waves developed. The storm waves filled our small catamaran boats, which sunk with our two-gradiometer magnetometers into the sea saltwater. We had now destroyed the $100,000 prototype equipment of Sheldon Breiner and Hewlett Packard. However, not to worry, because Sheldon Breiner left Hewlett-Packard and developed the world's main magnetometer company, that is world famous for geophysicists.

Unfortunately, we were now left with no geophysical equipment, so I had to develop a new research program. I decided that we would collect beach sediment samples that we could take back to the USGS laboratories in Menlo Park and examine them for gold and other heavy metals. My plan was that we would go out with the research boat and travel along the Kotzebue Sound shoreline (Fig. 3.2). I would take our small skiff with an outboard motor and travel near the beach and make many quick stops to collect beach sand samples. The small skiff could travel much faster than the slow research boat and this way I could go close to the shore and on and off the beach to quickly collect more samples. In the meantime, captain Frank would drive the research boat along the coast and later in the day I would take the skiff out to the research boat to travel back to our camp for the night in Deering. This seemed like a good plan, except sometimes at the end of the day, I could not find the research boat. Eventually, I would see the boat far offshore and when I got there I would find captain Frank drunk on the boat floor. I would then have to drive the boat back to our camp in Deering, Alaska.

Compared to my first adventure in Bering Strait with captain Frank, I had a small Russian adventure in Bering Strait at the end of the summer in 1967. I went on an oceanographic cruise with the University of Washington Thomas D Thompson research ship. We went to the Strait to measure the strong ocean currents and to study the bottom characteristics related to the currents. If you are not familiar with the Bering Strait, the western end of North America is Cape Prince of Wales, and 50 km (30 miles) across the Strait is the Russian peninsula of Chukotka (Fig. 3.2). In the middle of the Bering Strait lay the American Little Diomede Island and a mile away to the west, the Russian Big Diomede Island.

If you are young, you may not have heard that in 1967 the Cold War between USSR and the United States was raging (Fig. 3.1). This did not bother the American Alaska Natives or Russian Natives who freely went back and forth between the Diomede Islands. For us, it was important for our research work, because every time we entered Russian water west of the International Date Line we would immediately see Russian planes overhead or submarine periscopes above the water (Fig. 3.2). We could travel to the Russian side of Bering Strait up to 20 km (12 miles) offshore and take water samples and measure currents. We could not take any bottom samples on the Russian side, but we did anchor to the bottom when we were measuring ocean currents and sampling water. I avoided the problem of taking samples on the Russian side by looking at the anchor when it was pulled up and taking off any sediment samples that were on the anchor.

In the summer of 1968, there were more adventures related to my sampling for gold and other heavy metals as well as taking box cores to describe the animals living in the seafloor sediment (Fig. 3.3). The USGS was fortunate to use the research vessel Oceanographer that was operated by the US Coast and Geodetic Survey (now part of NOAA). For our sampling, we now used samplers that took large amounts of sediment up to about a half cubic meter (17 cubic feet), so that we avoided the problems of my original sampling off Nome Alaska in 1967. After our research expedition, these large samples were taken onshore where I hired Andy Peterson to pan gold from the samples. He was a 90-year-old man who was one of the most successful gold miners and still continued to mine gold up in the mountains near Nome Alaska.

Andy would stand 12 hours a day panning our samples for gold, but also continuously told stories about the gold rush days. For example, he had made and lost three fortunes because he owned a lot of property such as hotels in Nome, but fires destroyed each fortune. Another time he was returning with a dogsled in the winter and he and the sled fell into the river. He had to keep going a number of kilometers (miles) and arrived home just as he was becoming frozen solid. Eventually, as he was telling all his stories about the gold rush characters, I brought a tape recorder and recorded these fascinating stories of the history of the gold rush.

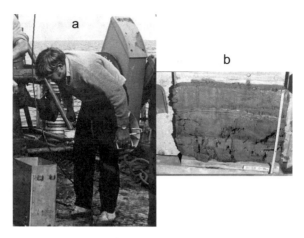

Fig. 3.3 **a** Photo of Hans Nelson sampling a box core on the research ship Oceanographer in 1968 in northern Bering Sea. The box corer frame is in the background and the lead weights are below the elbow of the person partially shown in the photo. The box corer frame is set on the seafloor and the lead weights push the box into the seafloor to fill the box with sediment. When the corer is picked up a shoe covers the bottom of the box so that the sediment is kept in the box while the frame is lifted through the water to the ship. The box of sediment can be sampled for the seafloor organisms, the layers of sediment and the chemistry of the sediment such as the carbon content to measure the biological productivity on the seafloor. **b** Box from box corer with one side removed to see the sediment layers. This box is not the one from Photo **a**, but is from a 1999 Cascadia core. Photo source for **a** is Hans Nelson and for **b** is Julia Gutierrez Pastor

Andy's story about his adventure with the dogsled reminds me of other interesting stories about Alaskan Native dog sleds. One of these involved a Native that I knew in Teller, Alaska (Fig. 3.2). John Burns, who worked for Alaska Fish and Game in Nome, told me that this Teller Native would come about 100 km (60 miles) to Nome in his dogsled to drink in the bars because Teller was a dry town with no alcohol. When the Nativae became drunk, they would bundle him up in his dogsled and the dogs would take him back to Teller while he was asleep in the sled.

Another less fortunate history, that I observed during my early years in Alaska, was how eventually most of the dogsled teams disappeared as Alaska Natives bought snowmobiles, which they called the iron dog. I would read stories in the Nome Nugget newspaper about someone who had lived many decades in Alaska and used dog sleds. Then later when traveling with a snowmobile, the person had frozen to death because the snowmobile had broken down mechanically or they were trapped in a storm. Taking long hunting trips into the country with dog sleds was a typical Native activity. Whereas a snowmobile could fail mechanically on a long trip, with dogsleds someone could survive by huddling together, or even as a worst resort like in Antarctica, they could eat the dogs.

Another classic hunting activity of the Alaskan Native Eskimos was to travel over the sea ice in the winter and hunt seals to eat. Unfortunately, this also was one of the most common ways that Eskimos suffered accidental death. My friend John Burns was involved in such an event. He and one of his good Native friends left Nome and traveled out onto the ice for seal hunting (Fig. 3.2). After they were far offshore, the ice they were on broke away from the shore and carried them away. The village of Nome Alaska gave them up for dead after a week when they did not return. However, by some good fortune, the ice sheet was blown back to the shore and they returned. This was like a miracle of returning from the dead for the townspeople and they proceeded to celebrate this miracle return for a week.

In 1969 there were more adventures when we continued our sampling for gold in the northern Bering Sea while using the Coast and Geodetic Survey ship Surveyor. The first of these adventures was when we stopped midway during the month long cruise and went ashore to have a beer because the ship again like all US ships did not allow alcohol. I got up in the morning after we arrived the night before and found that our USGS truck was missing from in front of our house. I looked down the street and could see the truck tire tracks going from side to side down the street. I followed the tire tracks down the street until I came to a church where the truck had been driven up onto the church steps. We then found out that one of the ships Hawaiian crewmen was in jail and had stolen the truck while drunk and driven it into the church.

That same night, just before a storm came in, some of the ship's young officers, who after a night of drinking, took a go-go dancer from a Nome bar out to the ship. Then after the storm came, no small boats could go back and forth between the shore and the go-go dancer was trapped on the ship and we USGS scientists were stranded in Nome for several days. It drove the captain crazy to have the go-go dancer trapped on the ship and finally even though the storm was continuing, he was determined to get her off the ship so a small boat took her to the shore. The boat then picked up the USGS scientists and we went out through the jetty and the high breaking waves. It was very dangerous and we nearly capsized, which had happened previously years before when some lives were lost. Then when we got to the big ship offshore, it was more stable, but the small boat was going up and down about 3 m (9 feet) so you had to jump for the ladder between the boats and then climb aboard the ship. All in all, this was a dangerous operation that never should have been undertaken, except for the captain's reasonable desire to get the go-go girl off the ship.

After the research expedition with the big ship, I wanted to do some other near shore sampling off Cape Prince of Wales. The reason for this was because of the fluorite and tin deposits in the rocks of Cape Prince of Wales along Bering Strait (Fig. 3.2). I wanted to see if these resource minerals might be concentrated in the ocean sediment by the strong currents heading toward Bering Strait. To do this, I flew with a bush pilot and landed at the Alaska Native village on Cape Prince of Wales. My plan was to rent a Native skin boat, called an omiak, which is built by stretching walrus skin over a wood frame.

When I arrived at the village, I could see no one or any sign of life. I went to the church parish in hopes that the preacher would have some information. I found that a few days before a booze bomber (a bush plane with alcohol) had arrived and the entire village had been drunk for several days and were still sleeping it off. I did manage to talk the preacher into taking his omiak out along the shore of Cape Prince of Wales. My sampling method was to take a small dredge sampler on the end of the rope, lower it over the side and pull up the sample by hand. This worked fine until the boat motor suddenly stopped. The strong currents rapidly swept us towards Russia on the other side of the Bering Strait. I was terrified that we would be taken prisoner and never heard from again. Finally, they got the motor started, because the Alaskan Natives are very good mechanics, and we got safely back to shore.

While in Nome Alaska during the summer of 1969, my main field assistant Dr. Bob Rowland, a Ph.D. geology student at University of California at Davis brought his wife Linda along, who also was getting a Ph.D. in sociology. She had a grand time finding out about the culture of Nome while working at a new café. The café owner claimed to be the previous cook for Trujillo, the dictator of the Dominican Republic, and he had to escape. In USA they say go West young man and Nome is the main town about as far west as you can go in North America. Consequently, a strange set of people collects in Nome. My father came to Nome to visit and talked with the owner of the café about cattle because my father was an agricultural major in university and raised some cattle on his farm. My father found all of the cook's stories to be knowledgeable and apparently he was truly who he said he was.

Another thing that Linda learned was that a group of mafia gangsters from Anchorage Alaska also had come to hide out in Nome. They were hiding out because they had killed someone in Anchorage and in the classic mafia manner had thrown the body weighed down with stones into the Turnagain Arm ocean area near Anchorage. Unfortunately, Turnagain Arm has a 20 m

(70-foot) tidal range and the body quickly appeared and they had to escape Anchorage. The gangsters fled to Nome where they found a comfortable life because they began the noble enterprise of bailing out Native Alaskan girls from the jail, who had been put there for pandering on the streets. Perhaps these young girls were mistakenly being promiscuous by carrying on the old Alaskan Native Eskimo culture of sharing wives. In the original Eskimo culture, it was considered gracious for a visitor to sleep with the host's wife of the family they were staying with. After the mafia gangsters bailed out the Natiave girls, they would live in the gangster house and provide a concubine for the gangsters.

Actually, some of the Nome townspeople were not much different than the gangsters. For example, the mayor, said to have a PhD from Columbia University, was known to be very devious and also apparently fled the lower 48 states to escape from some criminal activity. In fact, the editor of the Nome Nugget town newspaper had a feud for several months with the crooked mayor, and both walked around town with six guns ready to shoot each other.

I also had an experience with the mayor, who had let us use the town power plant building for our laboratory and gold panning. That summer our USGS truck died and I was given the task of announcing bids to buy our truck, because there was no way to take it back to California. I posted a notice at the Nome post office and then opened all the bids that came in. Once I found the highest bid, I sold the USGS truck to that person. The mayor had also made a bid that was lower, so I did not sell the truck to him. At the end of the summer when I went to thank the mayor for using the power station as our laboratory, he told me that I now had to pay him $5000 for the use of the building because I did not sell the truck to him. Obviously, this had been a common practice to sell to the mayor, because around his house he had an armada of military and other government vehicles that he apparently had obtained in this manner.

During the same time in the late 1960s, another Nome character managed to corner the northern Alaskan ivory market. Only Native Americans were supposed to be able to control the walrus ivory market. However, this person took his plane, filled it with alcohol, and then visited the Native villages of northern Alaska. He would give them the alcohol in return for their ivory and soon he had obtained most of the raw ivory. The Natives then had to carve his ivory, which he sold at a high profit to tourists in a number of shops that he owned. The last I heard of him, because he had to flee Nome, was that he had taken hundreds of thousands of dollars of his profits and bought oil leases in the North Slope of Alaska before the oil boom began. He probably ended up a multimillionaire because of his crooked ivory dealings.

Another of the interesting stories about Nome is, that through the years as I worked there, they gradually tried to modernize the town such as paving the streets and making concrete sidewalks. When they were making the concrete sidewalks, they had to tear up the old style of board sidewalks that you typically see in Western movies. When they did this in Nome, the townspeople got their gold pans and panned the dirt that had accumulated for years under the board sidewalks. They collected large amounts of gold because apparently during the gold rush days, the drunken miners would lose gold from their pockets and it would fall through the cracks in the board sidewalks into the dirt underneath.

An unfortunate adventure occurred after my 1969 summer research expedition. After our Alaskan cruise, we often got the cheap triangle fair to Hawaii to rest after the intense research cruise of the summer field season. In 1969, I stopped as soon as I arrived to body surf in the waves near Honolulu. While body surfing, someone broke into the car and stole everything including the only copy of my Ph.D. for publication and all my clothes. On a positive side, as a result of this robbery, I had one of the most exciting geologic experiences of my life. Because I had no clothes after the robbery and only my swimsuit, I called my good friend who then was Chief of the USGS Hawaiian Volcanic Observatory on the big Island of Hawaii. He invited me to stay with him, so I got on the plane in my swimsuit and when I got to his house, he provided me with clothes to wear.

At that time, the Kilauea volcano was actively erupting, so he took me into restricted areas to observe the volcanic eruption. We went to a place where lava had flowed out the day before and was still erupting in a crater. When we first arrived at the rim of the crater, lava was only bubbling deep down in the crater and I had to walk up to the edge to get a photo. I noticed that behind me there was a large crack in the rim. After I returned a few hours later, this piece of the rim on that I stood on previously, had collapsed into the molten pool of lava below. I had used up another one of my nine lives.

After leaving our first visit to the crater rim, we walked over a lava flow that had gone down the mountainside the day before. This fresh lava had a shiny silver surface, but in the cracks below the surface you could see red glowing lava. This crust sometimes broke as you walked on it, and you would suddenly drop about 30 cm (1 foot) or so down towards the hot lava. Also we watched in cracks below the surface where lava was flowing rapidly down the mountainside like a cascading mountain stream with large waves in the flowing lava. The lava was flowing in a lava tube as it traveled toward the sea.

Lava tubes are very common in volcanic flows and leave tunnels in lava after it has cooled. These are common in ancient lava flows of Hawaii and many other areas of the world. In some locations like Lava River Cave in the Newberry National Volcanic Monument near Bend, Oregon, you can walk for several kilometers through these dark lava tubes.

3.4   Mercury Pollution

After my gold studies in the late 1960s, I began new USGS studies on other heavy metal resources and pollution in Bering Sea. There were two aspects to these studies, one was to assess the potential for offshore platinum deposits as a resource and the other was the environmental assessment of mercury pollution. The studies of natural introduction and human-caused mercury pollution took place near Nome and in the Kuskokwim River drainage and coastal area. Natural platinum and mercury minerals were found as both lode and placer deposits in the Kuskokwim River area (Fig. 3.1b). Although there were onshore platinum deposits, we found no coastal or offshore platinum deposits.

The reason for my study of mercury pollution was that there was global concern about mercury pollution in the late 1960s and early 1970s, particularly in fish. This was one of the first signs of global change related to human activities. As described previously, the problem of mercury in fish was first noted in the smaller environments of lakes and rivers. There were specific inputs of mercury, especially from mining or sewage outfalls. The mercury in lakes also resulted from coal burning electrical plants, which introduced mercury as air fall directly into the lakes or was washed off the land into the lakes.

In both lake and marine environments, as described previously, there was biological concentration of mercury, so that large fish at the top of the food chain, which people mainly consumed, would have dangerous levels of mercury. These high-levels of mercury, which were first noted in lakes or rivers, soon were detected globally in ocean fish like swordfish, tuna and salmon. Whether in lakes, rivers or the ocean, the process of bio-concentration was the same. Low levels of mercury were eaten by the microscopic animals, which then were eaten by small fish, which would then be eaten by larger fish. However, with each step up the food chain, the levels of mercury became higher and higher until they reached dangerous concentrations for humans when eating too much fish.

The first location where I studied mercury pollution was in Nome where the gold rush had taken place (Figs. 3.1b and 3.2). Whenever gold was concentrated by hand panning, sluices or dredges, the gold flakes in the pan residue were amalgamated with liquid mercury metal, from which the gold could be recovered. I knew that when we panned samples from the modern Nome beach, we would find balls of liquid of mercury in the pan. This liquid mercury was left over from the gold rush days when miners would concentrate gold from their pans, but obviously some spilled into the modern beach sand. Thus, I wanted to assess what effect there was from this gold-rush mercury pollution. Chemists said that such liquid mercury could not exist since the gold rush days and would have been absorbed in the water. However, liquid mercury remained in the gold pans when we concentrated our beach samples by panning in the late 1960s.

We found that there was mercury pollution in the modern beach sand, but it was quickly diluted in concentration as soon as we looked offshore in sediment samples [3]. In this case, the engineers saying dilution is the answer to pollution was actually true. This dilution resulted because as soon as sediment with mercury was transported offshore, it was mixed with with large amounts of unpolluted sediment. The sedimet arrived in the offshore Nome area, because the third largest river in North America, the Yukon River introduces 60,000,000 tons of sediment into Norton Sound to the east of Nome (Figs. 3.1a and 3.2). Most of this sediment travels northwest past Nome in the Alaskan Coastal Current toward Bering Strait. This Yukon sediment mixes with the other offshore sediment and dilutes the Nome area sediment with mercury. Thus, there is no evidence of mercury pollution offshore from the Nome beach.

Later on because of my previous mercury studies, when dredging for gold was proposed to take place offshore from Nome, I was asked to evaluate if the gold mining would cause any mercury or pollution from the increased suspended sediment. In this case I could report that there would be no problem of mercury pollution, because of the transport of Yukon sediment past the Nome coast. Any mercury pollution would be completely diluted and any introduction of sediment from the dredging would be minor compared to the huge amount of Yukon sediment already flowing past this region. The bottom line is that careful environmental assessment studies should always be made to find out whether any proposed resource development will cause a pollution problem or not.

My second area of global interest about mercury pollution was to compare the effects from natural pollution of mercury deposits with those of human-caused mercury pollution. In other words, what were the natural contributions of mercury eroded from mineral deposits of cinnabar, the mineral containing mercury. The Kuskokwim River, which flowed into the Bering Sea from west-central Alaska, was the ideal place for this study (Fig. 3.1b). Three hundred sixty kilometers (225 miles) upstream from the coast there were natural cinnabar deposits along the river and the Red Devil mercury mine sites. I could start at these natural deposits and mine sites and progressively sample downstream from these locations to see what happened to the mercury contents in the river sediment downstream and also in the beaches near the river mouth.

I found that from the natural or mine sources there was a rapid decrease in the mercury content downstream from the sources [4]. Again this dilution was because sediment from other river tributary sources was being added and diluting the content of mercury from the natural cinnabar mineral grains and mining. The values for mercury from the natural erosion sources dropped downstream rapidly to unpolluted values in the upper river. In contrast, the polluted mercury values from the mining sources dropped to normal values further downstream in the middle river. The beaches near the Kuskokwim River mouth had no mercury pollution. Also there was no mercury pollution in the offshore sediment to affect the salmon fisheries in Bristol Bay (Fig. 3.2).

3.5   Encounters with Bear and Walrus

My logistics for the Kuskokwim River sampling resulted in several new Alaskan adventures (Fig. 3.1b). Because there were no roads in most of Alaska, I had to be transported by helicopter. This required that gasoline barrels had to be put out at locations along the river because helicopters only carry a small amount of fuel. To transport the gas, a small fixed wing plane was filled with 50-gallon (about 200 L) drums of gas. This certainly provided excitement because as you sat in the plane in front of the gas barrels you felt like you were in a flying bomb. In addition, the plane would land in remote runways sometimes overgrown with bushes so you could hardly tell that there was even a runway. You hoped that there wasn't something big in the bushes that the plane might crash into and explode.

One of our sites for gas supplies, and a place to stay for a night, was located at a camp in a recently abandoned mercury mine. After we deposited the fuel, we checked out the mine camp house that we were going to stay in. The house had the kitchen with the stove and refrigerator still there. As we went out of the house, immediately outside we found fresh bear dung that was still steaming and which had not been there when we went in the house. Both the kitchen appliances and the fresh bear dung were to become significant later, because it was obvious that a bear was lurking around the camp.

A few days later, we returned to the abandoned mining camp to stay for the night after we collected samples along the river with the helicopter. When we opened the door to the camp building we found that the bear had been inside. This was obvious because with a swipe of his paw he had batted off the solid steel refrigerator door and with another swipe of his paw he had batted off the steel oven door. But that was not all, because apparently mice were living in the walls between the gypsum wallboard and the outside wall. The bear had ripped out a lot of the plaster wallboard to expose and look for mice. To put it mildly we were very worried about sleeping there that night and I don't think the helicopter pilot or I slept for a minute.

The next day I began taking mercury samples around the mining camp. I looked up and saw that the bear was between the helicopter and me. This was not a good situation if I wanted to get to the helicopter and escape. I quietly waited and the bear slowly ambled away, and we left without the bear doing us any harm. In general, while doing my sampling I would be dropped off but many times had to go through the very dense brush to get to a sampling site. The only way possible to get through the brush was to walk through the trails that the bears had made. Of course I was very fearful while crawling along these trails and would make all the noise whistling and shouting that I could. Fortunately, the only thing I encountered was a lot of bear dung.

Another animal adventure while traveling with the helicopter along the beach was that we saw a dead walrus. I could not resist the temptation to try and obtain the one walrus tusk and the oosik bone (penis bone) on the walrus. Apparently, walruses die with their oosik up. I convinced the pilot to land and then found the only way for me to try to get the tusk and the oosik was to chop them out with a geologist rock hammer. This was not efficient and it took me a great deal of time chopping away and splattering myself with dead walrus, only to have no success at obtaining these trophies. Of course I did not smell very good and the helicopter pilot was not happy with my smell on the flight back to camp.

3.6   Assessing Hazards for Petroleum Development

In the mid to late 1970s, the USGS Menlo Park marine geology scientists, with NOAA-NEHRP funding, began a major new program of environmental assessment prior to oil exploration in Alaskan frontier basins. The United States government was required to make the assessment studies before any potential development of oil fields in these frontier basins off Alaska. I was assigned as the scientist in charge of studies for northern Bering Sea and the Norton Basin in the deep subsurface below Norton Sound (Fig. 3.2).

Our task was to assess any geologic hazards that may cause unwanted global change problems in our frontier basin area. These were hazards that might affect drilling offshore, the drilling platform site and potential development of pipelines across the seafloor from any oil well to where oil would be loaded on a ship [5]. In the Bering Sea, these hazards could be caused by active faults disrupting the seafloor, strong storm waves and storm surge to destroy offshore and onshore infrastructure on the shoreline, and gas-charged sediment that could cause an unstable seafloor. In addition, the shallow continental shelf of the northern Bering Sea had specific hazards caused by the strong currents, which could scour the bottom and potentially undermine infrastructure like pipelines and drilling platforms. Ice movement for seven months of the year also could destroy infrastructure of drilling platforms, scour the seafloor pipeline areas and present environmental problems if there was an oil spill.

The methods to study these potential geologic hazards for petroleum development included seismic profiling to look at the subsurface sediment layers and detect gas-charged sediment (Fig. 2.5). We utilized sidescan sonar to look at the seafloor surface characteristics such as active faults, gas escape features and sand wave movement. A new instrument called the geoprobe was developed, which could measure the near bottom currents, suspended sediment transport and storm wave pressure affecting the seafloor. In addition, we used geochemical studies to determine the type of gas in sediment and geotechnical studies to analyze the stability of the seafloor.

Some of these were new techniques that we developed for the special conditions in the northern Bering Sea. The most unique of these was the geoprobe developed by Drs. Cacchione and Drake [6]. At the time it was said that this was the most sophisticated instrument in the ocean except for nuclear submarines. It was the first instrument that could be set on the seafloor to take multiple current readings and directions near the seafloor, measure how much suspended sediment was transported and sense the wave pressure affecting the seafloor. It could sit on the seafloor and measure these characteristics for months, and if storms came by it would monitor conditions more often with all of these data stored in the seafloor instruments. Then the data could be retrieved when the geoprobe was called to release a float to the seafloor so the ship could pick it up and then shore laboratories could analyze the data.

We also developed special instruments to probe the seafloor offshore from the Yukon Delta area to assess the geotechnical characteristics (Fig. 3.1a) [7]. These instruments put probes into the seafloor to measure the conditions of how consolidated the sediment was and if the sediment was stable when extreme storm waves pounded the seafloor. These pioneering studies of sediment stability during extreme storms is even more relevant now as we see a greater number of more extreme storm waves in the offshore oil producing areas of the Gulf of Mexico. In addition, sediment cores were collected to take to the laboratory and measure the consolidated state of the sediment and its stability under strong wave and current conditions.

Sidescan sonar was utilized for some of the most important studies that we conducted in the Norton Sound and Basin area of northern Bering Sea (Fig. 3.2). The sidescan utilized sound waves sweeping out across the seafloor, which then made an electronic map like a photograph or hologram of the seafloor. By studying the sonographs of the Norton seafloor, we were able to assess ice gouging into the seafloor, active sediment waves moving across the seafloor and gas erupting from the seafloor, which made several meter diameter and deep pockmark holes in the seafloor. Underwater television and cameras were also used to observe the close-up details of these features. All of these features were hazards that could potentially cause oil spills from damage to drilling platforms and pipelines [5].

Dr. Keith Kvenvolden developed unique new methods to determine the composition of gas we found in sediment, after we discovered large subsurface areas that contained offshore gas bubbling to the seafloor to the south of Nome Alaska (Fig. 3.2). Keith used common paint cans, put core sediment with gas in the cans, inserted a steel needle into the can, and then analyzed the gas composition in real time at sea with a gas chromatograph [8]. After becoming interested in studying gas content in sediment on our northern Bering Sea cruises, he became a world famous expert on gas hydrates (frozen methane in seafloor sediment). This was significant because many gas hydrate experts estimate that they contain the world's greatest amount of hydrocarbons. Recently some Asian countries have begun preparation to try and mine these subsurface gas hydrates. In summary, our Bering Sea environmental assessment studies innovated many new techniques for marine geology and ways to prevent global change problems caused by oil spills.

Not only did we innovate new techniques in our marine geology studies in northern Bering Sea, but we also made many new discoveries about geologic hazards in these remote offshore areas that had little previous exploration. Some of our results were from typical types of studies such as mapping where recent faults have cut the seafloor [9]. Of course if significant oil resources were found in Norton Basin, active faults cutting the seafloor of Norton Sound could disrupt drilling platforms and pipeline infrastructure (Fig. 3.2). Other findings included mapping large areas of the seafloor that had linear ice gouges into the sediment from ice pressure ridges, gas pockmark holes caused by gas escape during extreme storms and active sediment waves that were moving across the seafloor toward Bering Strait, because of the strong currents. All of these newly discovered features also could be hazards to petroleum development infrastructure, just like more commonly known threats of active faults.

The long linear and curving gouges in the seafloor were caused by ice pressure ranges that developed as ice smashed to and fro and then the ice ridges were pushed across the seafloor by strong winds [10]. Sometimes ice movements of almost 100 km per day were observed by satellite. The result of all these ice pressure ridges cutting into the sea floor bottom was that over many years, the seafloor was completely gouged up in large areas of Norton Sound. Also in the same area, there were numerous holes or pockmarks that were several meters in diameter (Fig. 3.2). We found that methane trapped under a surface blanket of offshore Yukon mud, exploded out of the seafloor to form these pockmark holes when large storm waves of 6 m (20 feet) or more liquefied the bottom sediment and released the methane gas [7, 11].

Another discovery was that this area of shallow seafloor had a history of many strong storm surges. The strong winds of the occasional huge cyclonic storm's that struck the northern Bering Sea caused the water to pileup onshore and make a storm surge hazard. The town of Nome Alaska was certainly aware of this because it had been destroyed by storm surges several times during its hundred-year history. Another USGS coastal group studied the eastern end of Norton Sound and found that storm surge waves have piled water as high as 5 m or more than 16 feet above the normal sea level (Fig. 3.2) [12]. The water level in the shallow (<10 m, 33 feet) 200 by 400 km Norton Sound raised 50% within 24–48 hours because of the storm surge. When the extreme storm surge subsided, this huge amount of water trapped in the Norton Sound had to rush rapidly out of the sound as strong bottom return currents.

These unusually strong bottom currents transported large amounts of sediment 120 km across Norton Sound from the Yukon Delta region (Fig. 3.2). These seafloor deposits were up to 1 m (39 inches) thick and could be identified and easily mapped from sediment cores from the seafloor [13]. These pioneering methods now can be utilized globally to define the history of storm surges and the potential increase of extreme storms caused by global warming. Even in a small 1 m (39 inches) storm surge, these strong currents were found to carry over 70% of the entire suspended sediment transported in northern Bering Sea in one year, [6]. These currents also caused sediment waves to move across the seafloor as well as to scour into the seafloor. Thus another significant hazard was realized in the northern Bering Sea because these strong currents scouring into the seafloor and moving sediment waves also could destabilize drilling platforms or pipelines crossing the sea floor [10, 14]. In addition, the huge storm surge rise of sea level would be another major hazard for any petroleum industry infrastructure in harbor or shoreline areas.

3.7   Risks at Sea and on Shore

Our adventures related to the new USGS Alaskan program began before the ship even left for Alaska in 1976. Before beginning our scientific cruises, we had sea trials for all our new equipment in San Francisco Bay California. We tested our seismic equipment, which indicated where the bottom of San Francisco Bay was located. We then tested our piston corer, which allows deeper penetration into the sea bottom than gravity corers, which are just a tube with a weight dropped into the sea floor. The piston corer is triggered just above the seafloor and has a piston inside the barrel that slides up the corer to create a vacuum as the corer drops into the seafloor (Fig. 2.9a). Thus instead of getting a typical 2 m (6 feet) penetration from a gravity corer, with a piston corer you can get 6 m (18 feet) of penetration into the seafloor in sediment with sand layers or in a muddy seafloor, as much as 20 m (65 feet). So we lowered our $20,000 new piston corer to the bottom of San Francisco Bay to see how well it operated. However, San Francisco Bay has a thick soupy mud layer that is not solid, so that when our piston corer was triggered, it dropped below the soupy mud seafloor just like it was going through water. When the piston at the end of the cable in the corer tried to stop the corer going at full velocity, it broke the main cable of the ship and the corer continued going many meters below the seafloor and was lost on this first trial (Fig. 2.9a).

A similar accident happened in 1981 when I went on a coring cruise in the Mediterranean Sea and the Spanish made their first test of a piston corer on the Spanish research vessel Garcia del Cid. An accident with a piston corer is extremely dangerous because if the one metric ton or more heavy corer releases accidentally while hanging on the 1 cm (half inch) ship cable, the cable can break (Fig. 2.9a). If the cable breaks under this great tension, it whips wildly and if it hits a person it can cut them in half. The Garcia del Cid had a strange arrangement where the cable to the corer wound across the main deck through large pulleys and then went up and over a derrick frame and into the sea.

All of the Spanish scientists went and stood out on the deck to see their first trial with a piston corer (Fig. 2.9a). Because of my experience with piston coring, I made everyone clear the deck before we lowered the corer into the water. This was fortunate because the piston corer triggered accidentally in the seawater before hitting the seafloor. As a result, the pully holding the cable whipped across the deck, smashed into a box corer with 4 inch steel tubes and bent these tubes (Fig. 3.3a). If I had not cleared the deck, our entire scientific crew would have been badly injured or worse, so it was lucky I happened to be there for this Spanish piston corer trial.

After this digression about piston coring accidents, I return back to another different adventure that occurred in our 1976 sea trials offshore from San Francisco Bay. As we were returning towards San Francisco Bay about 16 km (10 miles) offshore, a huge sneaker or rogue wave came up behind the ship. Because it was not a stormy day, the door to the ship's galley (or dining room) was open. When the sneaker wave hit, water cascaded through the open galley door, swept everything off the stove and burned our cook with hot oil, but fortunately not too severely. The half-meter high water continued through the galley where we were eating and quite surprised by it, but no one was hurt.

A much worse event years later with a sneaker wave happened to some of my USGS scientist friends off southern California. There the sneaker wave turned their small research vessel upside down and they were able to break their way out of the upside down ship. They were close enough to swim to shore so that everybody survived. Because of being submerged upside down in the water, my surviving friend told me that he never felt good again while taking showers with water splashing over him.

Another similar miraculous survival happened to one of my graduate school friends Dr. Mike Laurs at Oregon State University. While he was on one of his first PhD research cruises in the Pacific, a wave swept him off the deck into the sea with no life jacket on. This normally would be a death sentence, because in stormy weather, it is very difficult to rescue anybody washed overboard. However in this case, another wave washed Mike back up on deck and saved his life. From my experiences and those of friends, it should be obvious that doing scientific research in the ocean is adventurous and can be dangerous.

While undertaking our 1976 cruise to northern Bering Sea, there also were logistics adventures. First, some of our equipment had to be sent to Nome by airfreight. However, in the early days of the late 1960s and middle 1970s, you never knew what would happen to your airfreight. The first priority of the airlines was to carry tourists because they made the most money from doing that. Whenever the airlines had extra space, they would take airfreight but might leave it anyplace in Alaska when there were more tourists to carry. Consequently, for many of my Alaskan cruises in the early days, my first task before the cruise was to call and find out where my airfreight was located in Alaska. One time when I could not find my freight in Nome Alaska, I called and the airline said we have over 50 thousand kilograms (110,000 lb) of airfreight waiting in Anchorage to go to the Arctic (Fig. 3.2). Sometimes my freight would be somewhere else in Alaska like Fairbanks. Then I would have to travel there and be sure that my freight got on the plane to Nome Alaska. Also if our flight with some of our freight stopped at another Alaskan airport before Nome, we would have to watch that they did not take our freight off before Nome.

When we took the new geoprobe with delicate electronic equipment to Nome I 1976, I remember standing with the two inventors and watching when they could not get the geoprobe equipment off the plane with the fork-lift. So finally they just hooked the cable to the pallet with the geoprobe and just pulled it out letting it drop about 4 m (12 feet) to the runway. The inventors were horrified and sure that it would never work for our cruise. Fortunately, they were able to repair it and we got incredible new data.

After the first use of the geoprobe in the northern Bering Sea, it was deployed in many other global areas. We had success in recovering the geoprobe in the Bering Sea, but this was not always true in other areas where it was used. There were times when fishermen caught the geoprobe in their nets and then refused to return it to the USGS unless a ransom of several million dollars was paid. Another famous event occurred when there was a problem retrieving the geoprobe, and an underwater robot was deployed to try and help retrieve it. This was back in the 1980s before sophisticated robots had been invented. An electrical cable connected the robot and a complicated set of hydraulic hoses to run it. An operator sat at a TV console on the ship and controlled the robot on the seafloor. While the robot was operating to retrieve the geoprobe, it suddenly malfunctioned, went flailing about, cut all the hydraulic lines, and completely self-destructed. All of this was captured on videotape and played as a hilarious event to watch. Of course it was not very funny for the owner of his only robot.

Our USGS Sea Sounder ship had other cruises for many months before ours, which was the last cruise of the season in October 1976. We had to meet the ship in Dutch Harbor, Alaska and needed to fly from Anchorage to Dutch Harbor in the central Aleutian Islands (Fig. 3.2). I will never forget the flight, because when we took off, the clouds were very low and only about 30 m (100 feet) above the sea. The flight had to stop at Sandpoint, about halfway between Dutch Harber and Anchorage. The pilots did not want to go above the clouds because they may not be able to land at Sand Point. So they flew below the clouds about 30 m (100 feet) over the ocean for four hours. Gradually the clouds got lower and lower and we were only about 7 m (20 feet) about the water looking at the whitecaps on the waves just below the plane. In Dutch Harbor, the runway is carved into the side of the island so that when we landed it looked like the wingtip was going to hit the cliff beside the plane. Flying in Alaska is always an adventure that you will see when I describe our return from the cruise in 1976.

Once we got to Dutch Harbor, the weather was bad as I mentioned before when describing our flight to Dutch Harbor. Because it was so rough in the harbor at first we could not get to the ship. So we spent several days waiting and of course visited the crazy bars in Dutch Harbor. These bars were full of huge fishermen mostly Norwegians. I have traveled all over the world teaching short courses with my late and best friend Dr. Tor Nilsen, who was 1.85 m tall (6 feet 5 inches) with flaming red hair like a Viking. I could always find him anywhere in the world, because his red head would always tower above everyone. However, this was not true in Dutch Harbor because most of the fishermen were about the same height and I could not find Tor in the bar. I thought it would be fun to dance with one of the girls in the bar, although Tor suggested that it might not be such a good idea. He was right, because while dancing a huge Norwegian fisherman, who was her boyfriend, came over and I quickly stopped dancing to avoid being obliterated by the boyfriend who stood more than a head taller than me.

Another adventure occurred when we were walking home from the bar one night with my two friends, Dr. Dave Cacchione and Dr. Dave Drake. Dave Drake suddenly disappeared as we were walking along the streets of Dutch Harbor, this abandoned World War II town that once housed 100,000 soldiers. We looked around and could not see him anywhere. Then we heard him shouting from below, because he had fallen into an open manhole that had never been repaired since World War II. Luckily, he was not hurt, and we got him out and went back to our hotel.

Finally, after the storm had subsided in Dutch Harbor, we again almost got hurt when the small boat from our ship in the harbor, came into shore to pick us up. This turned out to be another adventure, because one of the ships crewmen, Billy, always became completely drunk when he was on shore. After we got him into the boat, we started out of the harbor in a small channel with rocky outcrops along the side. However, Billy stood up and suddenly fell on top of the seaman driving the boat and we swerved sharply towards the rocky shoreline. Luckily, we turned just in time to avoid crashing.

This was not the only time Billy caused problems. During our same 1976 cruise, one time we had to stop mid-cruise to pick up more equipment in Nome. When we were returning from Nome, we arrived at the ship anchored out in the open sea offshore. Of course our small boat was smashing against the side of the ship as both moved up and down in the waves. Billy drunk as usual, fell out of small boat in between the two boats crashing together, but somehow was not crushed to death. This was the last straw for the captain who fired Billy and sent him off the ship back to California. Billy was very upset and threatened to send letter bombs to both captain Al and me the chief scientist. Fortunately, he disappeared and we never got letter bombs.

In addition to the problems with Billy, there was another problem with the ship in Dutch Harbor, because John the cook refused to leave for our month-long cruise without having any eggs. So we had to wait several days for the weather to clear for a plane to fly to Dutch Harbor with eggs from Anchorage (Fig. 3.2). After planning for a year and preparing all of our equipment such as the geoprobe, I was very frustrated that we were waiting for eggs and losing my ship time that was probably worth about $30,000 a day.

Finally the eggs arrived, and we went into a roaring storm in the southern Bering Sea. Most everyone, including me, did not have our sea legs and we all became seasick. However we had to keep preparing for the cruise and transiting to our northern Bering Sea study area. Thus I lay in my bunk describing tasks to my key scientific supervisors. In the meantime, my poor friend Tor was helping with the navigation, although he was very sick. He and the other navigator would take turns trying to keep records of the navigation while taking turns running outside to the rail and getting seasick.

While making stops in Nome Alaska, there were always the strange adventures in the bars like the Board of Trade. When we stopped for our first break in the 1976 cruise, we headed to Board of Trade to have a beer. My friend Dave arrived, and sat at the bar to have a beer. A drunk woman, next to him felt over on top of him just as another one of our scientists arrived, and then he said to Dave, who is your friend. Another full-time character, called Eddie the rat, essentially lived in the Board of Trade that was open 24 h a day. He survived by stealing drinks and food from trays as waiters walked by.

One time, late at night as we were finishing our beers, there was a table next to us with bar patrons who had been drinking all night. Just as we were about to leave, one of them fell over; he hit the next one, who fell over into the next one, and so on until they had all fallen over like dominoes and then the whole table fell on top of them. We thought this was a classic example for the closing of the Board of Trade. Also we sometimes brought our own entertainment to the bar because one of the scientists was a good guitar player and he would get up on the stage in the bar and lead all of us to sing folk songs. He had to share the stage with a strip tease dancer Miss flame, who danced with torches. Miss Flame had reached the end of the road in Nome, and I'm sure the end of her career, like many others who had fled as far west as they could in the United States.

Getting off the ship in 1976 turned out to be a greater adventure than we had getting on the ship. We arrived at the end of our cruise in late October when a bad fall storm was taking place. Thus it was not safe to take a small boat from our ship anchored offshore into Nome to catch our flights. We were concerned because many of us had the Western airlines triangle fare for $20, which would take us to Hawaii to warm up after the month of October in Alaska. However the storm continued for several days as all of the scientists waited to get off the ship.

We were especially frustrated because most of our wives had flown to Hawaii to meet us after we had been gone for a month working in the Bering Sea and our wives had been sitting for days in Hawaii waiting for us. In addition, the captain was very frustrated because after six months at sea in Alaska with no rest breaks for any of the crew, he wanted to begin the month long transit back to San Francisco Bay. Finally, no one could stand waiting anymore and the captain decided we would travel the 50 km (30 miles) from Nome to the more sheltered bay of Port Clarence, Alaska (Fig. 3.2). There we could offload into small boats and be taken ashore to the Coast Guard station that that had a good runway for planes which were ferried to Russia during World War II. In fact when I meet Dr. Lizitzen, a Bering Sea expert from Russia, he mentioned that the Nome area was familiar because he used to ferry planes from there to Russia.

When we got to Port Clarence, the storm was still raging and like my first adventure 10 years before in 1966, it still had rough seas for a small boat. One of the scientists and I decided we would try leaving our ship anchored in the bay and head several miles to the shoreline at the Coast Guard station.

We went in a zodiac boat and managed to surf into shore. However, it was extremely difficult to launch the zodiac back out towards the ship. We tried several times, but the zodiac would flip over in the surf. Finally, the zodiac did manage to get back to the ship. However only two people at a time could come ashore with the zodiac and each trip took about a half hour. Consequently, instead of quickly unloading the ship it took about four hours to gradually get everybody to the Coast Guard station.

We quickly could see that the Coast Guard station contained a very strange group of men who had been kept there a year without leave in this isolated location. The sailors sent to the Port Clarence station, were similar to those sent to Siberian gulags, essentially for bad behavior in the military elsewhere. We had a number of women scientists on the cruise and these trapped young sailors were beside themselves to see a woman for the first time in a year, so we had to protect our women scientists while they waited at the station.

To get to Nome Alaska from Port Clarence, I had chartered bush pilot planes to carry the scientists. These bush pilots had already flown for 14 h and they had expected to arrive in Port Clarence for a few minutes to load everybody, and then fly the 20 min to Nome, Alaska (Fig. 3.2). Because of the long time for everyone to get off the ship, the pilots had to wait another four hours before they could take off for Nome. Of course we had the storm still raging and we flew into storm clouds with heavy rain. As we flew along I saw the pilot continually taking a flashlight and looking at the wings. I asked them what he was doing and he said he had to watch for ice forming on the wings. Thus our plane would go up in our altitude, but then when ice began to form, the plane would have to drop down to have the ice melt off the wings.

I knew that there were the 1437 m (4700-foot) Kigluaik Mountains, which the plane had to climb over between Port Clarence and Nome. I also knew that the flight was supposed to be 20 min. Pretty soon a half hour had gone by and we had not arrived at Nome. Then an hour passed by and we still were not at Nome, and then another half hour. By this time, I was encountering one of the most terrifying times in my life, because I was sure we were lost and would crash into the mountain as had happened with many Alaskan bush pilots, including my late friend Bob Baker. Finally, after two hours we saw the lights of the Nome runway and were incredibly relieved. Our flight had taken two hours instead of 20 min and we had averaged a speed of 20 miles an hour over the ground. We felt extremely lucky to be alive and headed straight to the bars and celebrated our survival until the morning.

3.8   Gas in the Sea Bottom

Our 1976 discovery of gas in the sediment of the northern Bering Sea became an important topic of study for 1977. Gas charged sediment is unstable and can be a hazard to petroleum development infrastructure, which can result in widespread pollution from oil spills as well as gas leaking from sediment. As mentioned previously, methane gas was observed trapped close to the surface of Norton Sound [11]. The history in the Sound, because it is generally less than 20 m (60 feet) deep, was that during the ice age of lower sea levels, the area was covered with thick tundra organic debris. Then when sea level rose, the Yukon River mud deposited over and sealed this thick layer of organic debris. After the Yukon mud layer covered the tundra during the past 7000 years, the tundra organic debris decayed and created methane gas.

The gas could not escape through the Yukon mud layer, unless strong storm waves liquefied the overlying sediment. Our geotechnical studies in 1976 showed that the coarse silt-rich Yukon mud became liquefied when storm waves higher than 6 m (20 feet) were affecting the shallow bottom of Norton Sound (Fig. 3.2) [7]. Once this mud blanket became liquefied or unstable because of storm waves, the gas trapped below the mind could erupt and form the pockmarks or holes 3 m to 10 m (10 to 30 feet) wide in the seafloor. The discovery of pockmarks in Norton Sound in 1976 was one of the first observations of so called gas pockmarks, but soon they were found in the sea floor of many oceans around the globe [11]. The unstable bottom during storms and the several meter diameter holes of pockmarks in the seafloor creates a hazard for oil pipelines crossing the sea floor.

The other significant discovery in 1976 was a several hundred meter (about 1,000 foot) thick zone of gas-charged sediment that was observed in seismic profiles south of Nome (Fig. 3.2) [15]. Normally seismic profiles show a number of flat lines in the printed record, which are reflections from the underlying sediment layers. However when significant amounts of gas are in the sediment, all of these lines or reflections are obscured. In this approximately 100 km (62 mile) square area, that contained large quantities of gas, the seismic profiles also showed that gas plumes were streaming up through the overlying sediment and reaching the seafloor. In 1977, we undertook detailed studies of this gas-charged sediment that would be a geologic hazard to drilling platforms. Platforms could collapse by having legs sink into the unstable gassy sediment as has happened in the Gulf of Mexico during hurricanes. Collapsing platforms and breaking pipelines could have global implications from oil spills in Norton Basin because the strong currents could carry spilled oil far into the Arctic Sea.

Because of these hazards, we began detailed studies to determine the gas chemistry and to observe areas where gas was escaping at the seafloor. With underwater television cameras we could see gas bubbles streaming from the seafloor. Then the question was what type of gas was this, biogenic methane gas like we observed in Norton Sound, or deep underlying gas escaping from an oil reservoir or thermogenic gas. Our studies in 1977 indicated that there were thermogenic gases from a potential hydrocarbon reservoir. Of course this was exciting, because this would have shown that there might be petroleum reservoirs in the Norton Basin. However, the next year we returned and did further studies on the gas composition and found that almost all the escaping gas was CO_2 [8]. Thus, there was no indication of an underlying petroleum reservoir. Also a reservoir of carbon dioxide gas was not a resource in the Arctic, because there was no need to produce dry ice for refrigeration in the Arctic.

Another interesting chemical event that took place on this cruise was finding the rare mineral ikaite that forms only under certain conditions in ocean bottom mud. One of our cores from the Port Clarence mud had a large yellowish brown crystal mineral that filled the 10 cm (4 inch) width of the core barrel. As we watched after opening the core on the ship, the mineral evaporated into a pile of white dust. Some years later, I mentioned this to an English scientist Dr. Shearman, who happened to know about this odd mineral and he said you must have been the second person in the world to discover this mineral at that time [16]. Later other people also found this mineral in the seabed off Antarctica and Japan. After my discovery, Dr. Jim Bischoff, a geochemist at our USGS Menlo Park laboratory, ran experiments to explain the odd behavior of this mineral that was stable only at temperatures near freezing in bottom mud and evaporated into powder when it was exposed to warm air temperatures.

Our USGS environmental assessment cruises were a great opportunity for training young scientists. At this time in the 1970s, there was a pattern for recently graduated geology students to come and work as technicians at our USGS Menlo Park Marine Geology Branch. Typically, they would work for a year or two and then go on to graduate school to get a PhD. These technicians and interns had a tremendous chance to learn all of the marine geology techniques in the Alaskan region where many new discoveries were being made. Many of these students passing through our marine geology laboratory later became world-class marine geologists. Our Menlo Park Marine Geology Branch became a fountainhead for new scientists. In fact, one of my technicians eventually became the Associate Director of the USGS and another became the Director of the Smithsonian Natural History Museum in Washington. Consequently, our USGS environmental assessment studies were not only working on preventing human-caused global change, but also were providing a new generation of excellent scientists.

3.9   Ships Crew—The Good and the Bad

In addition to these studies of gas-charged sediment, in 1977 we continued to take many cores into the seafloor. Much of this coring took place in the shallow water of Norton Sound (Fig. 3.2). As in 1976, we had captain Al back, who was an incredible ship handler and would do all that he could to help us accomplish our scientific goals. Prior to working for us, Al was a pilot for the Panama Canal, so we had a great deal of experience of handling all types of ships in tight situations. Because of this, he did things no other captain would do, such as coring in 2 m (6 feet) of water close to the Yukon Delta shore.

The coring instrument that we had to use was called a vibracorer. It had a metal stand that rose 3 m (10 feet) above the sea floor so that it could support a large electric motor to vibrate the corer into the dense seafloor sediment off the Yukon Delta. Consequently, when we were in this very shallow water, the top of the vibrating corer was above the water and only a few meters away from the ship hull, because the electrical cable to the corer came from the ship.

It could take up to one hour for the vibracorer to drill into the sediment, while the ship bobbed up and down with the waves next to the top of the corer. One time while the drill was operating and captain Al was watching, I asked him why he seemed worried. He said that if he did not hold the ship at the right distance by constantly maneuvering engines to compensate for waves, currents and wind, the ship could cross over the top of the core frame and while going up and down with the waves it could punch a hole in the bottom of the ship. We were very lucky he had the skills and confidence to allow us to take these important cores close to the shore.

Other captains and crewmembers were not always so competent. In 1977, we had a terrible chief engineer for our cruise and it was disastrous for our electrical system on the ship. We had numerous blackouts with no electricity and we burned up our vibracorer electrical motor. Gradually because of all the power failures, one by one all of our electronic equipment failed. Eventually we had no seismic profiling system, sidescan sonar or navigation system that worked. Finally, not even any of the radios on board the ship worked, so with no equipment and no navigation we had to end the cruise. Another sign of the chief engineers' incompetence and habits, was one day when captain Al walked by the chief engineers' cabin, he noticed a terrible smell. He thought the sewer system had backed up, but he found that it was the chief engineer's room where he never showered, changed bed linen, or washed his clothes.

Another year, we had a chief engineer who was competent, but not a nice person. On the other hand John the cook, was excellent and got up at 4 am to make donuts and worked until finally the galley or kitchen was cleaned up by midnight. One night he was finishing the cleanup of the galley, while the chief engineer watched. The chief engineer was smoking, looked at John, and then threw his cigarette butt on the floor and ground it out with his foot on the clean floor. John was furious, grabbed a butcher knife and ran toward the chief engineer. Since I was there and observed all this, I ran to get between John and the chief engineer before he stabbed him. This is the second time this happened in my life because years earlier at the Crater Lake mess hall, I got between the cook with a knife and somebody he was mad at. Fortunately, in both cases, I convinced the cook to stop. However, an important lesson is don't ever mess with the cook.

Another dangerous adventure happened for me in the summer of 1977, when we were making a trip in the small boat from the Nome harbor out to our ship offshore. As we proceeded through the harbor entrance, the rope to tie up the boat was not secured and fell into the water, where it got caught in the spinning propeller of the boat. I was not watching and nor following one of the rules of the sea to never stand on top of a rope. Suddenly the rope wrapped around my ankle while it was being pulled into the boat propeller. If it had kept going I would have been pulled into the propeller and chewed up. Luckily just as I was about to be pulled off the boat, all the rope that had wrapped around the propeller killed the engine and I was not pulled into the propeller. However, to this day, 40 years later I still have a dark scar around my ankle from the rope burn as the rope was pulled around my leg.

3.10 Budget Cuts and Last Cruises

We returned to the northern Bering Sea for another two field seasons in 1978 and 1980 to continue our studies on geologic hazards, which could cause global changes from oil spill pollution. We wanted to further assess the hazards from storm surges and gas-charged sediment. This research required a larger vibracorer that could penetrate into the seafloor for 6 m (20 feet) compared to the smaller core that could only sample up to 2 m (6 feet) below the sea floor. We rented this corer from a private company, because it would have been extremely expensive for the government to buy one. We used the larger corer successfully in many locations in northern Bering Sea and then proceeded to Norton Sound where one of my most complicated Alaskan adventures occurred, as I will describe later (Fig. 3.2).

The new vibracorer sediment samples were examined with the previous 1960s samples to define geologic hazards as well as outline potential new mineral deposits. Some of these deposits contain a resource of rare earth minerals that that are valuable for technology manufacturing. In addition to all the geologic hazard and mineral information that we outlined, our data from environmental assessment cruises provided other valuable information about mammal populations that could be affected by potential oil spills. The discoveries about mammal populations resulted in new research in the 1980s that is described later (see Chap. 4 Sects. 4.5 and 4.6).

My last cruise to the northern Bering Sea was in 1980, and this was the last USGS scientific expedition in Alaskan waters. Starting in 1980 with President Reagan, large budget cuts began for natural science studies by United States government agencies. Thus, three fourths of the US offshore territory, which is off Alaska, has not had USGS marine geology studies since 1980. Also, no new marine geologists for the USGS have continued to carry on the research in this huge area that contains resources such as gold, fluorite and tin (e.g. [1, 17]). As we older scientists retire and die, we have not been replaced to increase knowledge in the vast Alaskan offshore area. Not only have we lost the research knowledge, but also the USGS marine geology group has lost all of its ocean-going ships that can work worldwide and now only has a few small coastal ships.

Unfortunately, the budget cuts for agencies like USGS have severely reduced natural science research at a time when natural and human-caused global change requires more study than ever before. For example during the 1970s our marine geology group in Menlo Park California had about 350 scientists and technicians plus two large research vessels. By 1998 when I left the survey our research group had been reduced to about 100 with no large research ships. After this time we had to use other university, government agency and private ships for our research. Our USGS marine research became much more limited to near shore areas and not in deep-water or polar areas. In fact later I will describe how the important USA goal of determining great earthquake history in the deep-sea off the Pacific Northwest was no longer possible (see Chap. 4 Sect. 4.13). I had to retire from the USGS to work with academic scientists and ships to accomplish our earthquake reseach that concerned a significant national need with trillions of dollars of real estate at stake in the Pacific Northwest and Northern California.

NOAA also lost many of its ships, so that at one point there were only two large vessels remaining for the Eastern United States Atlantic and Gulf of Mexico oceans and for the Pacific Ocean. This meant that each of the different types of oceanographic studies such as physical, chemical, biological and geological oceanography only had about three months of ship time a year to study these huge offshore areas that are much larger than the continental United States. Fortunately, the USA National Science Foundation UNOLS academic ships have continued to have funding and university marine geologists have been able to maintain their research compared to US federal government researchers.

3.11   A Ship with no Captain

When we went to Norton Sound in 1978, we found that the homogeneous silt was very dense and difficult to penetrate near the Yukon Delta (Figs. 3.1a and 3.2). However finally we vibrated the corer down 3 m (10 feet) into the seafloor. Unfortunately, when we tried to pull out the corer it would not budge. Finally, as we pulled on the ship's cable with even more force, the half-inch wire broke and screamed up the side of the ship fortunately missing the winch operator sitting on the side of the ship, because it would have cut him in half if it had hit him.

In the midst of this chaos, the captain came to the bridge to see what we could do. However, the 78-year old captain tripped over a chair, broke his hip, and went into diabetic shock while lying alive on the floor. With no captain, the crew seemed to fall apart and did not know what to do, but began sending routine messages on the radio. In the confusion, the mate pulled up the anchor into the side of the ship so that the anchor chain broke and the anchor fell to the seafloor. My main scientists also were not much help, because they were in the middle of a music practice session and did not seem to realize the severity of our condition losing the captain, the corer, and the ship's anchor within a matter of minutes.

When I saw the captain lying on the floor and realized he was in very bad medical condition, I called the U. S. Coast Guard, which the crew had not done. The Coast Guard was about 1000 km (620 miles) away in the Aleutian Islands and said they could not reach us for several days (Fig. 3.2). Consequently, we would have to transit the 24 h to Nome, and take the captain to the hospital there. We had to leave the captain lying on the floor of the bridge, because we did not dare move him with his broken bones.

Finally, when we got to Nome, the injured captain was put in a rescue litter basket and lowered onto a small boat to be brought to shore in Nome.

As the rescue boat was about to leave, I saw that the crew did not seem to know what to do, so I jumped onto the boat and went into Nome with the injured captain to arrange things at the hospital. The hospital quickly said they did not have the facilities to take care of his condition and he would have to be put on an airplane to Anchorage about 1000 km (600 miles) away (Fig. 3.2). This did not surprise me because my previous encounters with the hospital were for a sailor who thought he had legionnaires disease and a scientist who had bleeding ulcers. For the diseased sailor they prescribed that he drink a bottle of Jack Daniels and bed rest. For the ulcers, they suggested that the scientist swallow ice cubes, but I sent him home immediately on the plane.

For the captain, I arranged to get him on the next airliner to Anchorage. They took out some seats and put a hospital bed for him to lie on and arranged for a nurse to go with him. As I mentioned previously, planes were the only way to go throughout most of Alaska and one never knew what would be in the plane with you. I mentioned the time that we flew with four 50-gallon (200 L) drums of gasoline behind me. However another bush flight I took had a live reindeer tied up behind my seat. Nonetheless, we did get the captain safely to Anchorage where he was taken care of and survived, but was never able to go back to sea. This was sad because he started out his career as one of the last captains for a sailing clipper ship, and he was to be the captain for a new experimental sailing ship.

After dealing with the captain, I now had to deal with the problem of trying to retrieve the vibracorer that was stuck in the seafloor. We managed to find hardhat divers (i.e. with a metal helmet) that could go from Valdez in southeast Alaska where they were working on the Valdez oil tanker terminal at the end of the Alaskan pipeline from Prudhoe Bay oilfields. It was ironic that my cousin was the supervisor for the construction of the Valdez oil terminal. The divers, which were coming to Nome Alaska from Valdez, had been working on the terminal construction. The divers job had been to check welds in the casings for the terminal pier. These casings were 1.2 m (4 feet) in diameter and went 30 m (100 feet) below the seafloor. They would dive down inside these casings to check the welds and I cannot think of a more possible claustrophobic situation than being underwater in a tube about 30 m (100 feet) below the seafloor. It seemed they would be able to attack our problem of cutting the core tube at the seafloor, so that we could retrieve the corer back to the ship. We could replace the core tube lost on the seafloor and then keep working with the vibracorer.

After two days onshore in Nome, arranging for the captains' hospitalization and the divers, I went back to the ship to wait for the divers to arrive in Nome. However, unknown to me, there continued to be chaos with the ship officers after losing the captain. Our USGS Marine Facility in Menlo Park decided that the second mate of the ship should be named captain rather than the normal succession where the first mate would have become captain. The first mate was furious with me because he thought I had been in Nome arranging for the new captain, whereas I had nothing to do with making the second mate the captain. The first mate threatened to kill me, and I had to lock myself in my cabin for a day while he cooled off, but he remained very mad at me for the rest of the cruise. At the end of the cruise, I thought I would bring the first mate a peace offering and got him a bottle of whiskey as a gift. When I presented the bottle to him, he refused to take it because he said I was trying to have him lose his mates license. You can see he was quite paranoid, and still thought I was the one who kept him from being captain.

When the divers arrived we left Nome, to return to the site where the corer was stuck in the seafloor. The divers now had to take an underwater cutting torch and cut the core pipe at the seafloor so that we could retrieve the corer motor, hydraulic lines that ran to the corer and the metal superstructure that held the corer at the seafloor. This meant that the diver had to cut off the pipe, which was holding the entire superstructure that was suspended above him. If he did this wrong, everything would come crashing down on him. Also, the underwater cutting torch used 400 amps of electricity. If he touched the wrong thing he would be electrocuted. This was not a job for the faint of heart. Another observation was that you did not see many old hardhat divers from the oil industry. In fact, my neighbor lost her son, who was a diver working on North Sea oil platforms.

Fortunately, the divers worked safely, and we retrieved the vibracorer successfully. We made arrangements at the end of our cruise to meet the divers again in Anchorage to celebrate our achievement. We met the divers at Chilikoot Charlie's bar in Anchorage, a bar where a one-foot layer of peanut shells had accumulated in the typical fashion of an Alaskan bar. We scientists celebrated with a number of beers in this crazy bar. However, the divers offered us cocaine to celebrate with them, but we wisely refused this offer. I have often wondered how long these young divers survived their crazy and dangerous lives.

That same year, I also wonder what happened to the ship owner that I met in Homer, Alaska about 300 km south of Anchorage (Fig. 3.2). Prior to my 1978 cruise, I decided to visit Homer Alaska where many other Alaska USGS scientists had visited and thought was a beautiful spot. I drove down the Kenai Peninsula and my first activity was to try some salmon fishing for the first time in my life. The salmon were running up the river near Homer and the riverbanks were completely lined with fishermen. I joined the crowd and watched while everyone else caught fish after fish and I caught nothing. I'm not sure what the Alaska natives knew that I didn't know, since I never caught any salmon.

Thus after my ill-fated fishing expedition, I went into Homer and enjoyed walking around the town and the beaches where some homeless people were living during the summer. I met one person that survived by collecting the remnants of used shampoo bottles combining these remnants selling it as shampoo. This person lived on the beach in a small shack made of beach-combing junk.

In a Homer bar that night, I also met a ship captain who was drowning his sorrows with several drinks. As I talked to him, I found out that he had lost his ship in a strange way. He actually was a well-educated person, who had his degree from University of California at Berkeley. He decided that he would try to make his fortune fishing with new methods compared to most of the Alaskan fisherman. Most fishermen were very independent and used small boats with a crew of two or three people. This person thought he would try the methods used by most other countries where they would have a larger factory ship and small boats that would bring their catch to the ship. Consequently, he had a larger fishing boat than the typical fishing boats that worked out of Homer.

The captain and ship-owner had to make a repair on the bottom of his ship. The typical method to work on the bottom of a small fishing boat in Homer, was to drive the boat at high tide onto a wooden structure, let the tide go out and then the boat would be suspended in the air so that you could work on the bottom. This ship-owner drove his ship at high tide onto the structure and then when the tide went out, his bigger boat was too heavy for the structure and the structure collapsed leaving his boat lying on its side in the mud. Since I had no place to sleep, the ship-owner invited me to stay on his boat that was lying on its side and half filled with seawater. I took him up on his offer and went into the boat walking on the walls and then slept on one of the walls for the night. I left the unfortunate ship-owner in the morning to head to Nome, Alaska.

References

1. Nelson CH, Hopkins DM (1972) Sedimentary processes and distribution of particulate gold in northern Bering Sea: U.S. Geological Survey Professional Paper 689, 26 p
2. Miller KG, Wright JD, Browning JV, Kulpecz A, Kominz M, Naish TR, Cramer BS, Rosenthal Y, Peltier WR, Sosdian S (2012) High tide of the warm Pliocene: implications of global sea level for Antarctic deglaciation. Geology 40:407–410. https://doi.org/10.1130/G32869.1
3. Nelson CH, Pierce DW, Leong KW, Wang FH (1975) Mercury distribution in ancient and modern sediments of northeastern Bering Sea. Mar Geol 18:91–104
4. Nelson CH, Larsen BR, Jenne EA, Sorg DH (1977) Mercury dispersal from lode sources in the Kuskokwim River drainage, Alaska. Science 1198:820–824
5. Thor DR, Nelson CH (1979) A summary of interacting surficial geologic processes and potential hazards in Norton Basin, northern Bering Sea. In: Proceedings offshore technical conference, vol 3400. Houston, Texas, pp 377–381
6. Drake DE, Cacchione DA, Muench RD, Nelson CH (1980) Sediment transport in Norton Sound, Alaska. Mar Geol 36:97–126
7. Olsen EW, Clukey EC, Nelson CH (1982) Geotechnical characteristics of bottom sediment in northern Bering Sea. In: Nelson CH, Nio SD (eds) The Northeastern Bering Shelf: new perspectives of epicontinental shelf processes and depositional products. Geologie en Mijnbouw, vol 61, pp 91–103
8. Kvenvolden KA, Weliky K, Nelson CH, Des Marais DJ (1979) Submarine carbon dioxide seep in Norton Sound, Alaska. Science 205:1264–1266
9. Thor DR, Nelson CH (1978) Faulting, sediment instability, erosion and deposition hazards of Norton basin sea floor in environmental assessment of the Alaskan continental shelf, principal investigators reports, environmental research laboratory. Boulder, CO, NOAA, U.S. Dept. of Commerce, Aug-Oct, 1977, vol 3, pp 628–649
10. Larsen MC, Nelson CH, Thor DR (1979) Geologic implications and potential hazards of scour depression on the Bering Shelf, Alaska. Environ Geol 3:39–47
11. Nelson CH, Thor DR, Sandstrom MW, Kvenvolden KA (1979) Modern biogenic gas-generated craters (seafloor "pockmarks") on the Bering Shelf, Alaska. Geol Soc Am Bull 90:114–1152
12. Sallenger AJ (1983) Measurements of debris-line elevations and beach profiles following a major storm: Northern Bering Sea coast of Alaska, U.S. Geological Survey Open File 83–394, Menlo Park, California
13. Nelson CH (1982) Modern shallow-water graded sand layers from storm surges—a mimic of Bouma sequences and turbidite systems. J Sediment Petrol 52:537–545

14. Field NE, Nelson CH, Cacchione DA, Drake DE (1981) Migration of sand waves on the Bering Sea epicontinental shelf. Mar Geol 41:233–258
15. Holmes ML, Thor DR (1982) Distribution of gas-charged sediment in Norton Basin, northern Bering Sea. Geol Mijnbouw 61:72–79
16. Shearman DJ, Smth AJ (1985) Ikaite, the parent mineral of jarrowite-type pseudomorphs. Proc Geol Assoc 96:305–314
17. Nelson CH (1970) Potential development of heavy metal resources in northern Bering Sea. In: Proceedings 20th science conference in Alaska, 1969: Alaska Division American Association for Advancement of Science, pp 366–37

4

Global Lessons

4.1 Crater Lake and Volcanic Hazards

Previously, I mentioned in the introduction that volcanic eruptions can cause natural global changes and the Mount Mazama eruption forming Crater Lake was one of these global events (Fig. 2.4). The Mazama ash deposits are found on over one million square kilometers (620,000 mi^2) of western North America and in Greenland and Antarctic ice cores [1] (Zdanowicz et al. 1999). A similar sized historic eruption of Mount Tambora in Indonesia killed tens of thousands of people [73]. The Tambora eruption also resulted in sulfate aerosol in the upper atmosphere, which led to drastic weather changes in North America and Europe and caused food shortages. The following year, 1816, is known as the "Year without a summer" because of the major climate abnormalities. A previous Indonesian volcanic eruption of Mount Rinjani in 1257 AD appears to have caused the onset of a centuries-long Little Ice Age between the 14th and the nineteenth century.

Volcanologists fear that future volcanic eruptions the size of Mazama may cause catastrophic global effects including loss of life, and global disruption of transport and data transmission [73]. For example, prior to the 2020 pandemic, more than 100,000 commercial flights have taken place in the world every day, and vast volumes of data are transmitted via communication satellites. The potential failure of the satellite-based global positioning system (GPS), a fundamental resource for geographical positioning, would create unprecedented problems for transport systems. As a comparison, the 2010 eruption of the Icelandic volcano Eyjafjallajökull (500 times smaller

© The Author(s), under exclusive license to Springer Nature
Switzerland AG 2021
C. H. Nelson, *Witness To A Changing Earth*,
https://doi.org/10.1007/978-3-030-71811-4_4

than Mazama) resulted in volcanic ash spreading over part of Europe and the Atlantic Ocean [73]. Consequently, a large number of scheduled flights were suspended and caused estimated damages of over four billion Euros.

In 1979 I began the analysis of volcanic hazards under Crater Lake, because of the potential for future global changes from a volcanic eruption of Mount Mazama (Fig. 4.1). There were two ways of studying these hazards One was to examine the heat flow coming into the bottom of the lake, which would show if there were some molten lava or magma forming and coming closer to the lake floor for a potential eruption. The second method, which was the part that I was the chief scientist for, was to obtain seismic profiles to define the sediment layers under the lake floor (Fig. 4.2). This was to determine if layers were becoming gas charged from potential volcanic activity or if the flat lying layers were being bowed upward because of lava pushing up underneath the lake floor.

Another reason that we were doing the studies was to see if we could see gas trapped in water near the lake floor. I was concerned about trapped gas because the volcanic Lake Nyos in Cameron, Africa, had trapped gas in the bottom layer of the lake water. Suddenly in 1986 the water overturned and all of this CO_2 gas erupted and flowed at 100 km/hr (62 miles per hour) down

Fig. 4.1 The pontoon boat research vessel used for seismic profiling and sampling of the Crater Lake floor in 1979. The white circular tower at front receives signals for the boat's navigation system. Hans Nelson at the boat's stern provides a scale. The roof of the boat is where the small zodiac normally was tied, except for when the flying boat incident occurred (see discussion in Sect. 4.2). The caldera wall of the volcano surrounds the lake. The snow-filled gullies on the wall behind and to the right of the boat's white tower show where landslides occur on the wall. These landslides continue below the lake and feed sediment to the submerged aprons on the lake floor. See the arrow in Fig. 4.2 pointing to the nearsurface proximal high-amplitude reflectors and the aprons at the lake floor edge. See Fig. 4.5 for an example of these apron sediment layers. Photo source is U. S. Geological Survey

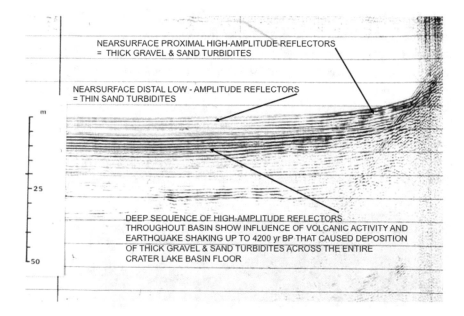

Fig. 4.2 High-resolution seismic profile from the Crater Lake caldera floor, which provides evidence for the lake floor history. The lower sequence of high-amplitude reflectors exhibits dark lines that cross the entire lake basin (see bottom arrow). These dark reflectors we interpret to be created by strong reflections from the coarse and thick sand and gravel layers deposited by large turbidity currents that traveled across the entire lake basin in the early history while volcanic eruptions were taking place on the lake floor. The correlation of strong reflections with sand and gravel layers is shown by the top right arrow pointing to strong near-surface reflections in Fig. 4.2 and the right arrow in Fig. 4.5 that points to thick sand and gravel sampled in the location of the strong reflections. The upper sequence of reflectors exhibits proximal high-amplitude seismic profile reflectors at the base of the caldera wall slope (see top right arrow Fig. 4.2). These reflectors evolve to faint lines of distal low-amplitude seismic profile reflectors towards the basin center (see middle arrow in Fig. 4.2). Core samples show that the proximal reflectors result from thick sand and gravel layers of debris flows (see right arrow in Fig. 4.5) and distal reflectors result from fewer and finer-grained turbidites (see left arrow in Fig. 4.5). Most important, all the reflectors are flat lying, which indicates that there is no present day lava pushing up into the lake floor and bowing up the seismic reflectors. The lack of bowed reflectors and disruption of reflectors by gas in the sediment suggest that there is no immediate danger of a volcanic eruption or poisonous gas escape at Crater Lake [2]. The location of the seismic profile is shown as line B to B' in Fig. 2.4. Figure source is Nelson et al. [1]

mountain valleys and killed 1,746 people because of suffocation from the layer of CO_2 that pushed out oxygen the people needed to breathe (https://en.wikipedia.org/wiki/Lake_Nyos_disaster).

Dr. Dave Williams led the study to see if there was increased heat flow in the Crater Lake floor. He obtained heat flow data from the lake floor sediment by dropping a corer several meters into the lake floor sediment and measuring the temperature gradient with depth below the lake floor. Dave found that there was a slight increase in temperature at some locations in the lake floor, which suggested there still was a small amount of volcanic heat affecting the lake floor (Fig. 4.3). This finding led to a series of studies a few years later by scientists from Oregon State University. These studies found blue pools of water in locations where the heat flow was slightly higher and a new type of bacteria that lived in the pools by reducing iron (Fig. 4.3 see

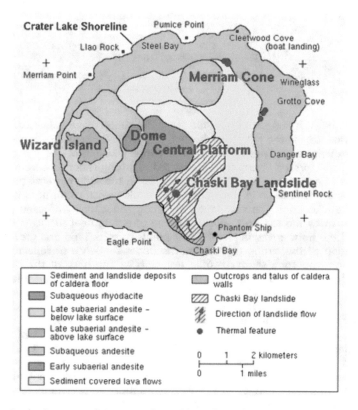

Fig. 4.3 Geologic map of Crater Lake caldera floor based on seismic profiles (see Fig. 4.2 example), dredged samples, manned submersible, and video from remotely operated vehicle. Map shows where subaqueous volcanic eruptions and deposits are exposed and their composition, where lake sediment layers deposited and the large Chaski landslide that went across the lake floor. The locations of thermal features or warmer lake water are shone with red dots. These thermal features are located on the ring fracture where where Mount Mazama collapsed during its cataclysmic eruption (Fig. 4.4). Figure is modified from Nelson et al. [3] and Bacon et al., 1997, USGS Open-File Report 97–487

large red dot in Chaski landslide). The Oregon State discovery in Crater Lake showed that oxidizing iron was another way that bacteria could live without oxygen.

My studies used the typical marine geology equipment of oceanographic research vessels. On our small pontoon boat we had the navigation equipment for location on the lake, high-resolution seismic profiling equipment, underwater television and photographic cameras (Fig. 4.1). We crisscrossed the lake taking many track lines of seismic profiles that showed the sediment and rock layers below the lake floor. We also took underwater television videos and bottom camera photos of features that we observed in our seismic profiles.

We obtained excellent results from our summer of 1979 research on the floor of Crater Lake. In spite of the slight heat flow Dave detected at the bottom of the lake, we determined that there were no imminent geologic hazards from renewed volcanic activity (Fig. 4.3) [2, 3]. No gas charged sediment was detected and no volcanic gas was trapped at the bottom of the lake. Thus there was no chance of a sudden escape of dangerous gas like there was at the volcanic lake in, Cameroon, Africa. Also there was no evidence of magma pushing up towards the lake floor because all of the sediment layers were flat and not bowed up (Fig. 4.2).

Besides the important volcanic hazards studies, we made new discoveries about the history of the lake floor after the Mount Mazama volcano collapsed and prior to the filling of the collapse caldera with Crater Lake water. Previous scientists had learned much about the cataclysmic eruption of Mazama, but little about the lake floor history after the collapse. With our new data we could map the deposits and volcanic structures under the surface of the Crater Lake caldera floor (Fig. 4.4). The volcanic structures began developing immediately after the cataclysmic volcanic eruption and collapse of Mount Mazama. Following, I will provide a summary of the Mount Mazama cataclysmic eruption as a background to understand our new findings about the lake floor history.

The eruption that destroyed Mount Mazama took place about 7,600 years ago, as shown in the year-by-year ice layers of Greenland (Zdanowicz et al. 1999). This cataclysmic eruption of Mount Mazama had two phases. First, 50 km^3 (31 mi^3) of volcanic ash and pumice (rocks with gas bubbles so pumice will float on water) were blown high into the atmosphere and fell down to cover over 1,000,000 km^2 (620,000 mi^2) of the Pacific Northwest, mainly the Columbia River drainage area [1]. Some of this Mazama ash washed off the land and eventually was shaken down in submarine landslides by Cascadia Subduction Zone earthquakes and deposited in seafloor sand layers of Astoria Channel (Fig. 2.8) [4, 5, 6]. Most of the ash and pumice remained on land,

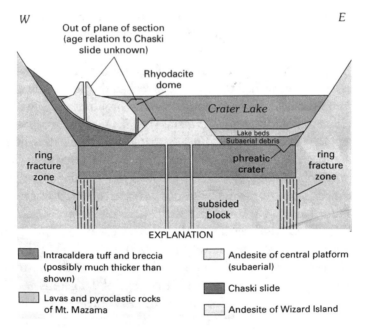

W

E

Fig. 4.4 Schematic geologic cross section of the caldera floor of Crater Lake showing relationship and sequence of formation of post-caldera volcanic features, subaerial debris layers, and lake sediment beds. Note that past (about 7,000 years ago?) phreatic (volcanic steam) explosion craters and modern day thermal features (Fig. 4.3) identify the location of Mount Mazama's collapse on a ring fracture, which created the collapse caldera in which Crater Lake is located (Fig. 4.4). The cross-section is not drawn to scale or as an exact cross section line. Figure source is U. S. Geological Survey marine and coastal geology program selected issues April 1995

where it is observed throughout the Pacific Northwest, particularly northeast of Crater Lake, where the yellow layer of Mazama ash is seen in road cuts. In the road cuts at Union Creek Oregon near Crater Lake, large charcoal logs from the forests destroyed by the volcanic eruption, are observed in the thick layers of ash along the highway.

The second phase of the mount Mazama cataclysmic eruption occurred after the first phase blew out and emptied 50 km³ (31 miles³) of magma underneath the mountaintop. Then the mountaintop began to collapse down while huge fiery avalanche deposits erupted in a ring fracture around the mountaintop (Fig. 4.4). These extremely destructive fiery avalanches traveled more than 160 km per hour (100 miles per hour) down the Mount Mazama valleys as far as 75 km (50 miles) down the Rogue River Valley [1]. Other known similar fiery avalanche events are those at Pompeii where people were frozen in place, or in 1902 where a fiery avalanche covered the town of

Martinique instantly killing about 30,000 people (https://en.wikipedia.org/wiki/Mount_Pelee). After the initial eruption and secondary fiery avalanche eruptions, the top mile of Mount Mazama collapsed along the ring fracture to form the 1200 m (4000 feet) deep hole or caldera that Crater Lake has now filled up halfway with water (Fig. 4.3, 4.4). At present high rates of precipitation, this only would have taken about 300 years to fill the lake to this level [3].

Our 1979 research shows what the lake floor looked like immediately after the collapse of Mount Mazama, and prior to the filling of Crater Lake with water. We see many huge landslides that slid down onto the caldera floor from the collapse of Mount Mazama (Fig. 4.4 see Chaski slide). We also observe the circular fault or ring fracture that the mountain collapsed along, which caused the circular shape of Crater Lake. This ring fracture is observed as a series of volcanic eruptions. These include Wizard Island that rises 700 feet above the lake surface, Merriam Cone that nearly rises to the surface of the lake, and the series of fumaroles (small volcanic vents) and phreatic explosion craters along the ring fracture (Fig. 4.3) [3]. After the 1980 eruption of Mount St. Helen's, which blew off the upper mile of this mountain as an explosive event, the floor of its large crater looked exactly like the caldera floor of Crater Lake just after the collapse. This was a nice confirmation of the subsurface seismic profile data that shows the original collapse caldera floor and then the lake floor sediments that deposit on top of the original floor (Fig. 4.3, 4.4).

The layers of lake sediment on top of the original collapse caldera basin floor also provide more of the lake history (Fig. 4.3, 4.4). The deep sediment layers have strong reflections indicating that course sand and gravel layers from caldera wall landslides extend across the whole lake floor (Figs. 4.2, 4.5) [3]. On top of these strong reflections, there are other strong reflections in submerged aprons along the edge of the lake [7]. These strong reflections of aprons evolve within a few hundred meters to weak reflections toward the center of the lake. From these observations, we interpret that the underlying coarse sand and gravel layers developed when larger landslides were shaking down from the caldera wall while large volcanic eruptions were still taking place to build Wizard Island and Merriam Cone (Fig. 4.3). An underwater ryodacite dome was the last active eruption feature and it pushed up nearly to the lake surface on the east side of Wizard Island (Fig. 4.3, 4.4). After this last volcanic activity smaller landslides went into the lake and deposited coarse sand and gravel in underwater base-of-slope aprons at the edge of the deep lake floor and then thinner, finer and fewer layers toward the center of the lake floor basin [7, 3] (Figs. 4.2 4.5).

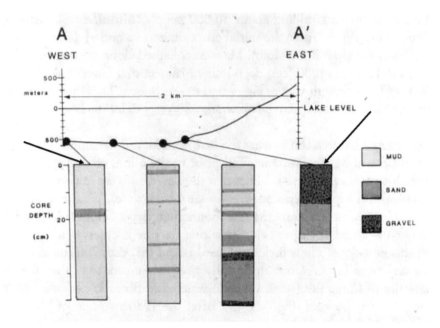

Fig. 4.5 This diagram of a transect of core samples across the Crater Lake basin floor shows the present-day evolution of the proximal to distal turbidites across the basin floor and provides evidence for the interpretation of the seismic profile reflections. The arrow to the right shows the thick sand and gravel layers in the aprons at the base of the caldera slope, which cause nearsurface proximal high-amplitude seismic profile reflections (see top right arrow in Fig. 4.2). The arrow to the left in this figure shows the thinner, fewer, finer sediment size turbidites toward the basin center, which cause the nearsurface distal low-amplitude seismic profile reflections (see the middle arrow in Fig. 4.2). The location of the seismic profile is shown as line A to A' in Fig. 2.4. Figure Figure source is Nelson et al. [1]

Previously, I obtained a sediment age of about 4200 years on top of the dome (Figs. 4.3, 4.4). Combining this knowledge with our new seismic profile data, we interpret that significant volcanic activity took place from the time of the collapse about 7600 years ago until 4200 years ago [3]. After 4200 years, when volcanic eruptions were not shaking Mount Mazama, the normal sediment processes of random small landslides sliding down the caldera wall and into the water have occurred (Fig. 4.5). These landslides deposit thick sand and gravel aprons at the base of the caldera wall, after which turbidity currents flows transport the sand out across the flat lake floor (Figs. 4.2, 4.5) [7].

4.2   A Flying Boat at Crater Lake

Our work at Crater Lake during the summer of 1979 had complex logistics, just like there were in Alaska. Our first task was to get our research boat to the lake. There is no road from the rim down to the Crater Lake shore and all equipment must be taken down a narrow trail along the Crater Lake wall that is 30 to 45 degrees steep and rises 300 m (1,000 feet) above the lake level. Consequently our pontoon boat had to be lowered on a cable and slid over the snow during the spring (Fig. 4.1). Then all the other equipment had to be loaded on a small trailer and taken with a small tractor down the lake trail to where the boat was docked. We also carried many items on our backs. Luckily, we got everything down to the boat, but when we were taking our equipment up at the end of our research, the trailer with our corer rolled off the trail and everything fell several hundred feet down into the lake never to be seen again.

We almost didn't survive our first day of work on the Crater Lake in late June 1979. The road, which must be taken to the other side of the lake from the Park Service headquarter did not have all the snow plowed to open the road to the lake trailhead until the Fourth of July weekend. Our first day, the road had been partially plowed to the lake trail, but was not open to the public. A bad snowstorm came up while we were down on the lake and when we tried to return back to the park headquarters, large snowdrifts blocked us. Fortunately we had four-wheel-drive vehicles with winches and we were able to gradually work our way through the snowdrifts and return home. If we had been trapped behind the snowdrifts it would have been dangerous with no shelter, water or food.

Our first activity on Crater Lake was to set up navigation stations around the lake. These instruments had to be set up at several locations around the lake edge, which was 300 m (1,000 feet) below the caldera walls. The loose volcanic debris often had small landslides and the rocks and small pebbles would come screaming down to the lake edge while making a whirring sound like bullets over our heads. Even with our hardhats, if we had been hit, the hats would not have made a difference and it would have been like being shot by a bullet.

Our most famous lake adventure was called the flying boat incident and it became world-famous with marine geologists. Sometimes, while working on the lake we stopped at the dock on Wizard Island and sat on the island and had lunch (Figs. 2.4, 4.3). This was the same thing that tourists taking the boat trip around the lake also did. One noon we docked our small pontoon boat on one side of the dock and while we were eating lunch sitting up on a lava flow, the tourist boat tied up on the dock on the other side of our boat. We always carried a small zodiac boat on top of the awning of our pontoon boat, because each day at the lake trail boat dock, we had to tie our pontoon boat up offshore to a buoy and then take the zodiac to shore (Fig. 4.1).

Of course this zodiac normally was tied to the top of our boat, except someone forgot to do that on this particular day. Thus, as we looked up from lunch when a gust of wind came along, we saw the boat fly through the air and come down and hit the only single person, a lady still sitting on the tourist boat at the Wizard Island boat dock. The impact of the flying zodiac boat knocked her unconscious. Fortunately, she soon revived and seemed to have no bad injury, to our great relief. However, a flurry of paperwork, and worry that we would be sued followed, but in the end this never happened. The main result was that the story of this event rapidly traveled around the marine geology world. A few months later that year, I was at a scientific meeting in Europe and overheard geologists talking about the flying boat incident on Crater Lake.

Another strange lake adventure occurred one day when we were towing the bottom camera, which had a large steel frame protecting the camera. We were towing the camera slowly above the lake floor, but apparently the frame caught on a rock outcrop. Suddenly while I was driving the boat, I saw the wake in front of the boat rather than behind the boat in the normal manner as the boat traveled forward. What had happened, was that after the camera caught on the rock outcrop, we kept going forward stretching the cable like a rubber band. When the cable could be stretched no further, it pulled us back like a rubber band so that the boat was rapidly going backward instead of forward, until the cable was not stretched anymore.

Another incident was not so pleasant for Matt, one of the scientists on our Crater Lake expedition. To power all of our equipment on our pontoon boat, we had a small electric generator driven by a gasoline engine (Fig. 4.1). There were many beautiful sunny days while working on the lake, and this happened to be one of them, so Matt was wearing shorts. Without looking, he backed into the red-hot muffler of the gasoline engine and severely burned the back of his leg. Of course at Crater Lake, there were no doctors or hospitals and all we could do was to take Matt to the Park Service nurse on duty. She applied ointment and a large bandage to the back of Matt's upper leg. Unfortunately, the bandage healed into the skin on the back of Matt's leg, and then we were faced with the problem of how to get this bandage off of Matt's leg. The nurse seemed to have no solution, so we decided to give Matt a bottle of Jack Daniels to drink and then we ripped off the bandage while

Matt screamed bloody murder. I am sure none of us, or Matt ever forgot this and he probably has a scar to this day.

4.3   Politics and Science

In 1982 I moved to Washington DC where I worked in the USGS Directors Office and represented the Marine Geology program. This was an eye opening experience to see how the administration worked at the national headquarters and also experience meetings with other government agencies. Some of these meetings were quite interesting, because my former PhD thesis advisor Dr. John Byrne was the Director for NOAA at that time and I saw him at several inter-agency meetings. One of the most striking things at these meetings was how little knowledge there was in Washington about the huge offshore territory that USA has in Alaska (Fig. 3.1b). The USA offshore area is larger than the onshore area and three fourths of this offshore area is found off Alaska in the Pacific Ocean, Bering Sea, Chukchi Sea and Beaufort Sea. There was little knowledge of these offshore areas from the politicians and agency administrators in Washington.

This lack of knowledge and interest continues today in Washington even though national security, resources and the most critical global change problems are important in these Alaskan offshore areas. As mentioned previously, there have been no USA research studies by USGS in the vast Arctic offshore area since 1980, although some university studies have continued. However, these academic studies typically are detailed research projects and the large USGS comprehensive studies of global change have not taken place.

With all of the interest in USA national security and military spending it is unfathomable that even today there is little concern about Alaskan areas adjacent to Russia. In particular, the United States only has two old ocean going icebreakers and they have had to be rescued several times in the Arctic by Russian and Canadian icebreakers (https://en.wikipedia.org/wiki/List_of_icebreakers). Compared to the United States, the Russians have 27 ocean going icebreakers with 11 new planned or under construction and many that are nuclear powered https://www.google.com/search?client=firefox-b-e&q=number+of+russian+ice+breakers. The Russians already have built a floating nuclear power station for the Arctic and in December, 2019 it began operation off Chukotka across the Bering Strait from USA (Fig. 3.2) https://en.wikipedia.org/wiki/Russian_floating_nuclear_power_station. Russia plans to make 7 more such stations and no one has knowledge of what will happen with the nuclear waste.

The USA also has little knowledge of the previous large amounts of nuclear wastes that the Russians dumped in the Arctic after the Cold War. We do know that the circulation of the Arctic ice gyre is transporting these wastes to the northern Alaska coast. The USGS had a program to study this global change of nuclear waste contamination, but it was abandoned in the mid-1990s because funding for this USGS project was cancelled.

At present, the Russians are investing huge amounts of money and resources in development of their Arctic infrastructure. In essence, the United States has ceded the Arctic Ocean area to Canadian, Russian and Scandinavian countries because we have few polar ships and little or no military or federal research presence in this area. This seems like a huge lack of security, resource and trading interests, because with global warming the Arctic passage for ships will become a reality and the development of many new resources will be possible.

The lack of interest in the Arctic area and USGS marine geology program began under President Reagan when his Secretary of Interior was Richard Watt. Richard Watt was a creationist and certainly did not have much interest in the USGS, which of course studied earth history that has taken place over billions of years, not just 6000 biblical years. Not only that, one of Watts signature programs at the time was to create the Minerals Management Service (MMS), which really was a regulatory agency mainly for offshore petroleum leases. However when he created the agency he added the USGS marine geology research program, as part of the MMS, but this was a mismatch of missions.

Immediately many other government agencies and particularly state geologists complained that USGS marine geology should not be part of MMS, but to no avail. Finally Dr. Michael Halbouty, a famous Texas oilman took on the task of trying to get marine geology back into the USGS Geologic Division. Dr. Halbouty, not only was an oilman, but he also was an excellent research geologist that was actively working with USGS geologists. Because of his knowledge of the USGS, he went to visit Watt and requested that the marine geology group be put back in the USGS Geologic Division. Watt, the Secretary of Interior, just threw Dr. Halbouty out of his office and refused to listen to his request. However, Dr. Halbouty was in President Reagan's so-called kitchen cabinet that advised him on many issues. Dr. Halabouty went directly to President Reagan's Oval Office, and the next day USGS marine geology was saved and put back in the USGS Geologic Division where it rightfully belonged.

In 1983 I visited China and represented the USGS director's office to explore the possibilities of joint research projects. This was very early when

China had just opened up to outside visitors from United States. When we landed in Shanghai, a scientific colleague of mine made a comment that with the heavily polluted sky, everything in China was gray. It immediately was evident how much China contributed to global change pollution. Fortunately, now the Chinese government is actively trying to reduce their air, water and soil pollution. In contrast, the populist USA government in 2017 began eliminating climate change research and pollution regulations. These are important lessons showing how critical it is that there is political will to address global change problems.

There were several minor adventures on my trip to China, such as soldiers with guns everywhere and guarding the hotels. In 1983 you could only go on government protocol visits and had to be guided everywhere, although they did put you alone on trains or planes and a translator would pick you up amongst the thousands of people at the end of a trip. This worked fine until I arrived at the airport in Guangzhou (formerly Canton). No one was at the airport to meet me, and the Chinese had given me no contact or hotel information, as was usual throughout my visist to China. I was stranded for two hours with no idea what to do, and with no money since the Chinese provided all the travel, lodging and meal expenses. Finally, two Chinese geologists came up laughing hilariously, because they had picked up the wrong blond geologist and taken him to Guangzhou before they realized their mistake. Fortunately, the rest of my visit was quite pleasant staying at a beautiful hotel by a lake.

4.4 Tiller Whales

Global warming is affecting the Arctic more rapidly than anywhere else (Fig. 4.6) (see also Chapter 5, Sect. 5.6). As a result, the marine mammals there are feeling the effects, particularly the Pacific walrus. Some of my previous research on gray whale and walrus feeding now provides a baseline to monitor these global warming changes that are affecting these mammals [8]. Prior to my environmental assessment studies on the Bering continental shelf, scientists were aware that the Chirikov Basin in northern Bering Sea was a major feeding area for Pacific gray whales and walruses (Fig. 3.2). In the early 1980s, we began reviewing our high-resolution sidescan records of the Bering seafloor to see if we could find evidence from these mammals feeding in this area (Figs. 4.7, 4.8a, d). The biologists knew that gray whales migrated to feed mainly in Alaska during the summer. Along the California and Pacific Northwest coast each year people observed the gray whales going south in

Fig. 4.6 Map showing northern Bering Sea water temperatures in 2019 that are from 2.8 to nearly 5.5 °C (5 to 10 degrees °F) above average in the main Pacific walrus habitat area of the Arctic. This temperature increase resulted in significant melting of sea ice in 2019 (e.g. Figure 4.11). Figure source is https://edition.cnn.com ›2019/06/28›alaska-sea-ice-wxc

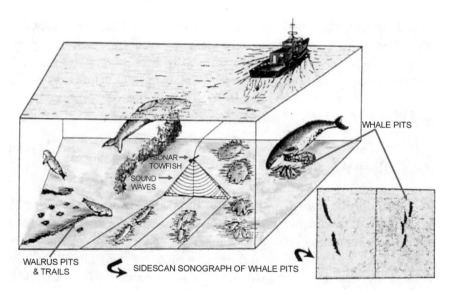

Fig. 4.7 Schematic diagram showing sidescan sonar method used to document Pacific gray whale and Pacific walrus feeding pits and trails on the northern Bering seafloor. A boat tows a torpedo-shaped transducer known as a towfish above the seafloor. The towfish sends multiple sound waves that scan the seafloor and then the towfish receives return signals, which vary in intensity depending on the shape of the seafloor structures. The shipboard recorder translates the signals into a sonograph (see Fig. 4.8 example), which looks like a photograph and shows the whale and walrus feeding pit and trail structures. Figure source is Nelson and Johnson [8]

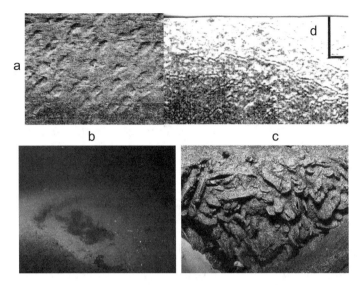

Fig. 4.8 **a** Sidescan sonar sonograph showing feeding pits made by Pacific gray whales in the seafloor of northern Bering Sea. The smaller newly formed fresh whale feeding pits are about the two-meter (6 feet) size of a whale mouth. The older larger pits have been enlarged by ocean bottom currents and are oriented obliquely across the sonograph in the direction of the current flow. **b** Seafloor photograph of a freshly formed whale feeding pit that is approximately 2.5 m (7.2 feet) long by 1.5 m (4 feet) wide and 10 cm (4 inches) deep. Photograph source Larry Martin of LGL Ecological Research Associates, Inc. **c** Photograph of the mucus-lined tubular burrows built by numerous half-centimeter (quarter inch) amphipods in the Chirikov Basin feeding area of the gray whales (Fig. 3.2). The network of tubes binds loose sediment into a mat that gray whales take bites out of leaving behind pits that are seen in parts **a** and **b** of this figure. **d** Sidescan sonar sonograph of the Pacific walrus feeding trails in the seafloor of northern Bering Sea. Vertical and horizontal scale bars equal 10 meters. Figure source is Nelson and Johnson [8]

the fall along the coast and then north in the summer. It was also known, that when gray whales proceeded north in the summer, they had 30% less weight than they had in the fall returning, again showing evidence for their main feeding areas in Alaska [8] (https://en.wikipedia.org/wiki/Gray_whale# Feeding).

The first main feeding region for the gray whales is in the Aleutian Islands area and then proceeding northward along the continental margin of the southern Bering Sea where the world's largest submarine canyons are located. Each of these canyons has intense upwelling of nutrient rich deep Pacific water, which results in a high productivity of zooplankton (small microscopic animals). The gray whales are known to be very omnivorous baleen filter feeders and can feed both by straining out the zooplankton from the water as well as from the seafloor sediment. After feeding in the southern Bering

Sea, whales proceed into the northern Bering Sea once the ice had receded [8](https://en.wikipedia.org/wiki/Gray_whale#Feeding).

In the early 1980s, the marine mammals' scientists had been observing and counting the gray whales while they fed in the Chirikov Basin area between St. Lawrence Island and Bering Strait (Fig. 3.2). During my 1960 and 1970s I sampled this area with box corers that push a box into the seafloor and collect a 0.15 m squared (1.6 sqft) by 0.5 m (1.5 feet) deep section of the seafloor sediment (Fig. 3.3). In the box cores I found that this area was dominated by amphipods [9]. When I would bring up a box core, the top would be covered with masses of these small up to 1 cm (0.5 inch) orange amphipods (Fig. 4.8c). These crustaceans lived in mucus burrows in the upper few centimeters of the seafloor. They were so dense in this area that the upper few centimeters of the seafloor was like a mucous mat. The Chirikov Basin area was the ideal sediment for them to live in, because it was a sheet of fine sand that had been deposited as the shoreline transgressed and regressed across this shallow continental shelf area during the ice ages [10]. Because of the strong currents, a mud blanket did not settle over this ancient sand deposit. From the stomach contents of gray whales, it was known that these amphipods were the main food of the whales in the Chirikov Basin.

At the time before our studies in the northern Bering Sea, National Geographic cartoons showed gray whales plowing into the seafloor sediment with their mouths. If this was the way they fed, they would have rapidly sand papered their mouths and died. However, when we began studying our detailed high-resolution sidescan sonar records, we found that the seafloor in the Chirikov Basin was covered with small pits about 2 m (6.5 feet) long and 1/2 m (1.6 feet) wide (Figs. 4.7,4.8a, b) [8]. We realized that this was the same size as one side of a whale mouth and interpreted that these were whale-feeding pits. We interpreted that the whales fed by extracting their tongue and sucking in a mouth sized bite of the amphipod mat and then straining out the sediment mat through their baleen and collecting the amphipods to eat (Figs. 4.8c, 4.9).

When they strained out the amphipods, they collected mainly the adult-sized amphipods and the small juvenile amphipods were spewed out with the sediment (Figs. 4.7, 4.8c). This surface sediment of the seafloor was extremely rich in nutrients so that the result was the young juveniles were cast like seeds into a fertilized seafloor. Consequently, when I described our new theory of gray whale feeding in a February 1987 Scientific American article, I titled it gray whales and walrus, tillers of the sea floor [8]. Essentially the whales were like farmers harvesting the adult amphipod population each year and then returning the next year to harvest the new adult amphipod crop from the

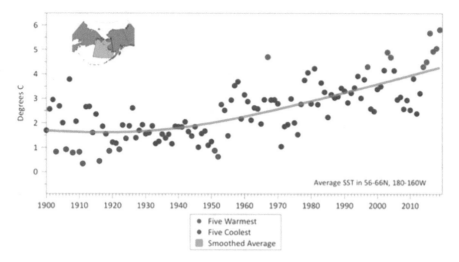

Fig. 4.9 May to June Bering Sea surface temperatures from 1900 to 2019 showing that the highest average temperatures have occurred since 2015. Figures source is NOAA ERSSTvS processed by ACCAP (Alaska Center for Climate Assessment and Policy)

previous year's juvenile amphipods. It is ironic that my first year working in the Bering Sea in 1967, while flying across the Chirikov Basin, we saw many whales with brown water plumes trailing behind them. I thought the plumes whale poop, but I know now that these were whale sediment feeding plumes (Fig. 3.2).

We also could determine a great deal more about the feeding habits of the gray whales. We could see that there were areas with a high density of whale feeding pits and on the fringes there would be a few pits leading into the area with a high density of pits (Fig. 4.8a) [11]. When we compared this with the data from a University of Alaska scientist, we found that the area with a high density of pits had a total organic carbon content of 250 g (0.6 lbs) of carbon or more per square meter [9]. This is quite high because outside the Arctic seafloor, a typical average is about 50 g (0.1 lb) /m^2 (10 sqft).

The bottom sediment in the Arctic seas, like Bering Sea, can reach as much as 2000 g (4.4 lbs)/m^2 (10 sqft), and this is why both the gray whales and walruses feed in these areas [9]. We observed, that when the amount of organic carbon was less than 250 g (0.6 lbs) /m^2 (10 sqft), there was a low density of whale feeding pits. We interpreted that the whale feeding strategy was to test the bottom with random pits and then when a high density of amphipods was found such as shown in Fig. 4.8c, whales concentrated their feeding in this area and left a high density of feeding pits (Fig. 4.8a) [8]. This whale behavior was logical, because if there were not an extremely high quantity of amphipod food, the whales would spend too much energy for the amount of food and would lose weight. The whales needed the weight gain to exist without eating during the winter when they traveled to and from Baja California, where the young gray whales were born (https://en.wikipedia.org/wiki/Gray_whale#Feeding).

We also could determine how much the whales were eating and how many whales the northern Bering Sea feeding area could support. Normally for mammals like deer or bears, it is hard to determine what population can be supported in a feeding area, because it is difficult to measure how much grass or berries these forest animals eat. In contrast, we could measure the total area of whale feeding pits made in a year and we knew the quantity of amphipods that were eaten each year by the whales in the northern Bering Sea feeding area [11]. At the time we did our studies in the 1980s, the gray whale population was estimated to be about 19,000 whales and recently it has been estimated that the population has reached as many as 27,000 (https://en.wikipedia.org/wiki/Gray_whaleWikipedia). From our studies of the amount of food available, we estimated that this food supply in the northern Bering Sea could support as many as 22,000 whales, although this is not their only feeding area.

These populations of gray whales were an example of a remarkable recovery over a century, because at the beginning of the twentieth century after the intense whaling activities of the nineteenth century, only a few hundred gray whales remained (https://en.wikipedia.org/wiki/Gray_whale#Feeding). However, when the gray whales were protected from whaling, the population rebounded to its natural full amount by the end of the twentieth century. Another indication that the Bering Sea had reached its full carrying capacity for the gray whale, was that whales began feeding in new Chukchi Sea areas and eating things they had not previously eaten.

Now there are questions about how global change may be affecting the gray whale population. A Philip Cousteau documentary on gray whales reported that there has been a severe drop in the population of gray whales at the beginning of the twenty-first century. Based on our previous studies, we could help find out what was causing the drop in the number of gray whales. We could make a new assessment of the area of seafloor whale feeding pits and the amount of fixed carbon eaten from amphipods. Is the warming of the sea from global climate change changing the amount of amphipods available to eat, or are different species migrating into the amphipod area because of warming seas? We have provided methods that could be used to find out what global change is doing to the gray whale population. However, with the severe funding cuts for government natural science and climate change denial, the political will is not there to do these important studies to help us monitor the effect of climate change on gray whales or how to keep a sustainable planet and animal species from extinction under threat from global warming.

It became evident while President Reagan was in office in the 1980s that the political will towards natural science research had changed. During his term, it was decided that federal government funding and the numbers of scientists for research should be cut and that scientific research should be privatized. Consequently, when some new studies for the gray whales were proposed after our findings, neither the USGS or NOAA were funded for these new studies. Instead private industry was contracted to do these new studies in northern Bering Sea. The first year the private contractor went to do the study, it was a complete failure because of their lack of knowledge of the science and logistics in working in the remote northern Bering Sea.

Not only that, when we conducted our studies in northern Bering Sea our USGS federal funding for science was about $100,000 per year. In contrast, the science funding for the private contractor was $700,000 per year, almost 10 times more than it had been for our similar USGS research. The government was not saving money, nor was it achieving any research results. In addition, the private contractor was from Canada and not a US private company, so that the profits and employment did not even help the USA. In the second year of the private research cruise to northern Bering Sea, I sent one of my research scientists and some of our equipment to assure that there were research results for the summer. In sum, cutting USA research budgets was not saving money, creating jobs or achieving research results.

Without our research funding for northern Bering Sea cruises, we began some local studies for the gray whales that did not require expensive Arctic expeditions. It was known by marine mammals scientists that the male gray whales would migrate northward to feed ahead of the females with young. The females migrate more slowly and stop in some areas with limited food along the way to teach there newborn young to feed (https://en.wikipedia.org/wiki/Gray_whale#Feeding). Thus, these females and a few other whales could be found along the Pacific Northwest Coast and one of the locations was along Vancouver Island off Canada. We went to study a small group of gray whales along the coast of Vancouver Island.

We could see there were whale-feeding pits on the seafloor from initial feeding and then during our studies we observed that whales were straining zooplankton from the water along the west coast of Vancouver Island. This also turned out to be an adventure because we conducted some of our studies from a small zodiac boat. While observing the whales, one time a gray whale came immediately next to the boat so that we could almost touch it. The most memorable thing from this event was the terrible bad breath of the whale when it exhaled through its blow hole. Fortunately, that was the only adverse part of this experience from the friendly whale, which did not flip its tail or overturn the zodiac that we were in.

Also while studying gray whales off Vancouver Island, we observed a pod of Orca killer whales. Years later I had a similar experience when serving as a lecturer on an Antarctic ecological tour. Part of that experience was to be a zodiac driver for tourists as we made landings onshore. One time near the Antarctic Peninsula, a group of Orca killer whales was sighted and we went to this location to see them in our zodiac. When we got there we could see the killer whales were circling a humpback whale and it's calf and moving in for the kill. When our and other zodiacs arrived, killer whales dispersed but after that, the humpback and it's young calf continued to stay near our zodiac, maybe thanking us for driving away killer whales and saving her young baby.

4.5   Disappearing Walrus

At the same time, while studying our sidescan sonar records for the whale feeding features, we also came in contact with Dr. John Oliver's marine mammals group at the California State University, Moss Landing Marine Science Center. They were examining Pacific walrus feeding in the northern Bering Sea and knew that the walrus had two feeding methods (e.g. [12]). The most common feeding method is to swim along the seafloor with their tusks sliding along the sediment while the tentacles on their flat nose sense where near-surface clam siphons are sticking above the seafloor (Fig. 4.7). Once the tentacles sense a siphon, the walrus chop off the siphons and eat them. Clams can regenerate the siphons, so the walrus have a continuing crop for food.

The second walrus feeding method is for deep clams like Maya that can be a foot below the seafloor. In this case, their tongues pump up and down like a piston and jet water to make a pit and excavate the clams so that they can put their mouths around the clam and suck the clam bodies out of their shell (Fig. 4.7) [12]. With these feeding methods, the stomach content of walruses shows they never have shells but just clam bodies in their stomach. If they had a feeding method where they accumulated shells, they would have quite a lot of indigestion.

I learned another interesting fact about why walrus did not accumulate a stomach full of shells. The skull of the walrus has a circular piston-like shape, which allows the walrus to jet water and excavate the clams. Thus, when pulling in the tongue, the walrus can extract the claims from their shells. I witnessed another example of this feeding method one time when I visited Marine World in San Diego. There were some walruses that had been there for many years and when fish were thrown to feed them they caught them in their mouths and swallowed them. However, there were some young walruses that had only been there a short time. When the fish were thrown, they put their mouth on them and tried to suck them in like they were a clam and the young walrus would just suck the fish in half and then eat them.

With this knowledge of the walrus feeding habits, we reviewed our high-resolution sidescan records from northern Bering Sea where the seafloor had not been chewed up by the gray whale feeding. There were meandering walrus feeding trails along the seafloor for hundreds of meters (yards) and small circular walrus feeding pits of a few tens of centimeters (one-half foot) wide (Fig. 4.8d) [8]. These seafloor features matched with the known feeding methods of the walrus. Using these trails in the seafloor, we could map the walrus feeding areas in the northern Bering seafloor [12].

These walrus feeding features covered the seafloor because of the lifestyle of the walrus living along the edge of the sea ice [8]. The sea ice gives them a platform to rest, a place to raise young and a habitat to live on. The edge of the sea ice at the end of the summer is in the Chukchi Sea far north of Bering Strait (Fig. 3.2). As the cold weather of the fall begins, the ice grows southward until finally in late winter the ice edge is at the shelf edge in southern Bering Sea. Consequently, the advance and retreat of the ice edge is like a conveyor belt where the Pacific walrus habitat and feeding area travels back and forth across the entire walrus feeding areas in the Bering and Chukchi seas each year (Fig. 3.2). This feeding area has resulted in a maximum population estimated to be from 201,000 to 246,020 walrus at the time we were making our studies in the 1980s [13].

With global warming, this crucial habitat of the migrating ice edge has changed for the Pacific Walrus. The ice edge now often does not reach the southern end of the Bering shelf as it did prior to global warming (e.g. Fig. 4.10). This reduced ice area results in less feeding area for the walrus population because they traverse a smaller part of the seafloor for feeding since the sea ice does not grow as far south as previously. Consequently, the population has become stressed, as shown by the large numbers of walrus now hauling out of the water and crowding onto the Diomede Islands, Sledge Island, St. Lawrence Island in northern Bering Sea and other shorelines of Alaska and Russia (Figs. 3.2, 4.11) (Netflix Our Planet, [14]. In 2014, an estimated 35,000 walruses were observed on the beaches just in northwest Alaska (Fig. 3.1 A pink box and Chukchi Sea) [15]. A result of this crowding onto

Fig. 4.10 Comparison of Bering Sea sea ice (shown in purple color) on March 17, 2013 with the same day in 2019, which shows an example of the drastic loss of normal sea ice cover in northern Bering Sea habitat area of the Pacific walrus. This extreme ice loss does not happen every year, but Arctic sea ice reductions have been averaging 13% per decade and this limiting factor has resulted in a loss of half of the walrus population since the 1980s. Source of sea ice figures is from NOAA AMSR2 (coastwatch.noaa.gov/cw/satellite-data-products/sea-ice/amsr2-sea-ice) processed by ACCAP (Alaska Center for Climate Assessment and Policy)

land can be walrus deaths by smoothering and falling. A horrific example has been shown by the Netflix Our Planet series where hundreds of walrus have climbed cliffs to avoid crowding and then fallen off cliffs to their death when possibly chased by polar bears (Fig. 4.11).

The walrus population now appears to have been reduced from its previous peak population of about 250,000 because of the smaller feeding and living area of the sea ice. For example, recent studies have estimated that the walrus population is 129,000, which is half of the previous population in the 1980s [13, 16, 17]. This indicates that global warming already has significantly

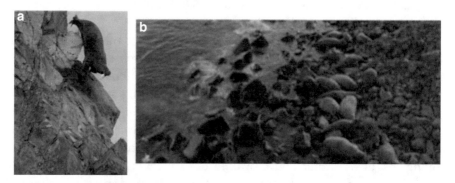

Fig. 4.11 Photos of Pacific walrus **a** falling off cliffs and (B) to their deaths below. Deaths result from overcrowding on islands because of the extreme loss of the walrus normal sea ice habitat (see ice loss example in Fig. 4.10). Photo source is from Netflix, [14], Our Planet Episode 2, Walrus. See Chap. 4 Section 4.5 for explanation

affected the Pacific Walrus population and lifestyle. Fortunately, there is some evidence that the walrus are beginning to adapt to hauling out on land and traveling increased distances to feed compared to living on the ice edge. Because of this possible adaption, MacCracken et al. [16] believe that the walrus will survive, although in significantly reduced numbers.

There are many articles published showing the potential global warming effects of sea ice changes affecting polar bear populations. However, the global warming change has already caused drastic changes for walrus populations (Fig. 4.11). Previously this global warming effect on walrus has not been publicized much in the press, perhaps because walrus may seem a bit ugly and not as supposedly charming as polar bears, which hunt and eat people. However, the recent Netflix series, Episode 2 of Our Planet has helped to alert the public to the walrus population reduction caused by global warming.

4.6   Mediterranean Sea

I learned several new lessons about both natural and human-caused global change after I became a co-chief scientist for a large cooperative USGS and Spanish project. This project studied both the Mediterranean and Atlantic offshore margins of Spain. The program lasted from 1984 to 1990 and included about 35 scientists from the United States and Spain. Our research covered the entire sediment systems from their river source to their final deep-sea depositional sites. This study included river sediment input, physical oceanography, chemical oceanography and marine geology of the entire continental margin consisting of shelf, slope and rise with submarine fans (Figs. 2.7, 4.12, 4.13) [18].

Our research began in 1984 in association with Columbia University, Lamont Doherty Marine Geology Institute. We utilized a deep-towed sidescan sonar to map the multiple Ebro submarine canyons and channels, Valencia Valley and Valencia Fan at the end of Valencia Valley in the Mediterranean Sea off the northeastern coast of Spain (Figs. 2.5, 4.12, 4.13). This sidescan study was done with Dr. Bill Ryan, who with others made one of the outstanding discoveries of natural global change in the twentieth century. They proved that about 6 to 5.3 million years ago, the Mediterranean Sea dried up and became a 4000 m (12,000-feet) deep hole that was then covered by a thick salt layer from the drying up of the Mediterranean Sea [19]. This had many implications for the history of the Mediterranean Sea, because deep canyons eroded into the dried up basin walls (e.g. Figure 4.12).

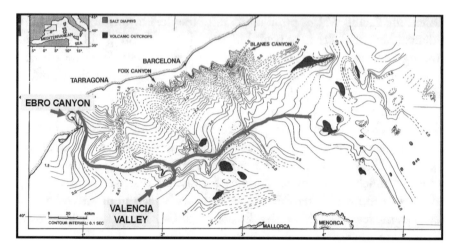

Fig. 4.12 Map shows the Mediterranean Sea as a dried up basin about 5 million years ago. The red line traces the axis of the enlarged, deep canyon that was eroded by the Ebro River at this time when the basin was a 3,700 m (12,000 feet) deep hole. The river canyon drained into the Valencia Valley. This land surface of 5 million years ago was determined by seismic profiling where sound waves reflected off the distinct salt layer of the dried up seafloor. Figure source is Escutia and Maldonado [20]

My wife, Dr. Carlota Escutia, used seismic profiles to map these deep subsurface canyons and river systems in the dried up Mediterranean seafloor (Fig. 4.12) [20]. Her study was important because during our later USA Spanish project, we discovered that the large gullied canyons of the modern Ebro margin are located in the area that is underlain by the ancient Ebro Canyon of the dried up seafloor (Fig. 4.13) [68]. The gullied canyons have been created by multiple submarine landslides in the thick, unstable marine clay that fills this deep underlying ancient Ebro Canyon (Fig. 4.13) [74].

The Mediterranean dried up because natural tectonic forces closed the Strait of Gibraltar and there was little or no Atlantic water input to the Mediterranean Sea. Without the input of Atlantic water, the Mediterranean Sea dried up in a few hundred years. When the Strait of Gibraltar opened up again, the Atlantic seawater rushed in and rapidly filled up the Mediterranean and deposited the thick unstable clay that filled the ancient Ebro Canyon (Figs. 4.12, 4.13) [21]. Dr. Ryan and others determined this history with the Deep Sea Drilling Program [22]. When they drilled through the deposits under the present day Mediterranean seafloor, they encountered the evaporated seawater salt overlying reddish oxidized gravels from when the Mediterranean seafloor was exposed to air [19]. Then on top of the salt from the dried up seafloor, they found 5 million year old deep-water clay that

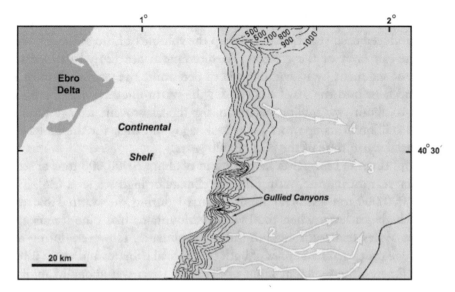

Fig. 4.13 Map shows the seafloor offshore from the modern Ebro Delta in the western Mediterranean Sea. Four narrow canyons cross the continental slope and feed into channel levee complexes formed by turbidity currents that drain into the modern deep-sea channel that overlies the ancient terrestrial Valencia Valley of 5 million years ago. These channels developed during different sea level lowstands of the Pleistocene ice ages. Channel 4 is the oldest, channel 1 is the youngest, and the three larger gullied canyons formed where the continental slope is underlain by the ancient large terrestrial Ebro River Canyon of 5 million years ago (see Fig. 4.12). Figure source is Nelson et al. [68]

was deposited when the Mediterranean rapidly filled up shortly after tectonic forces opened up the Strait of Gibraltar.

With Dr. Ryan, we also studied the submarine Var Fan off the Var River that passes through Nice France. In contrast to the ancient natural global change of the dried up Mediterranean Sea, the Var system became famous because of a 1979 submarine landslide that apparently was human-caused (see Sect. 4.11). In 1984, we were one of the first groups to do an academic study of the submarine landslide and the deposits on the sea floor. We observed huge waves of gravel and sand on deep-sea floor of the Var Canyon mouth [23]. These sediment waves were deposited by the high-speed turbidity currents (meters or yards-per-second), which were triggered by the submarine landslide.

Similar to the human-caused 1979 landslide, a major surprise of human-caused global change occurred from our USGS-Spanish study of the Ebro River system (Fig. 4.13) [18]. One of the tasks that I undertook was to determine the history of the Ebro River sediment load and its transport into the deep Mediterranean Sea. I calculated the amount of sediment that the Ebro River produced each year by summing up the volume of Ebro sediment overlying the salt layer of the dried up Mediterranean Sea [24]. I divided the volume of sediment by its age in years to determine the yearly transport of Ebro sediment into the sea. I did this for the warm Pliocene time from five million to about two million years ago, for the Pleistocene ice age time of about 2 million years ago to 10,000 years ago, and then for the Holocene post-glacial warm time of the past 10,000 years.

I found that the Ebro River had an input of about 6,000,000 tons of sediment per year during the warm Pliocene climate of high sea level [24]. The same 6,000,000-ton sediment input occurred during the warm Holocene time of high sea level, when the Pyrenees Mountains had pine forests and the Ebro Valley had oak forests of the warm climates. However, during the Pleistocene time, that had cycles of both cold glacial climates and warm interglacial climates, the sediment input of the Ebro River was about 15 million tons a year average for both these cold and warm climates. During the glacial period of cold climates, the pollen records showed that the Pyrenees Mountain pine forests and the Ebro Valley oak forests disappeared. The pollen records also showed that when the Romans entered Spain 2,000 years ago, there was a change to pollen that was mainly from grains, grapes and other agricultural crops instead of the pine and oak forests [24]. The extensive cutting of forests continued during the time of shipbuilding for the Spanish Armada. By the time of the beginning of the twentieth century, the Ebro River sediment input reached 22 million tons of sediment per year.

This was an astounding finding for me, because it showed that humans were affecting the Earth's environment in a way equal to the ice-age climate. In other words, the effect of humans changing the vegetation in Spain equaled what the ice ages did, but in much less time. It was my first recognition of the Anthropocene (e.g. [25]), that human-caused global change was now the controlling geological force for the Earth. Another human effect was that during Franco's government in Spain, a major program began to build dams on most of the rivers to provide hydroelectric power. Consequently, a number of dams were placed on the Ebro River, which caused most of the river sediment to be trapped behind these dams. As a result, instead of the normal 6 million tons of sediment input from the Ebro River to the Mediterranean Sea, there now is only about 100,000 tons of sediment input [24]. This loss of sediment input is causing the Ebro Delta to erode back at an average of 1 cm (0.3 inch) per year compared to the average growth into the sea of 4 cm per year with the normal 6 million tons of sediment input. The long-term affect is that Spain's richest farmland of the Ebro Delta will gradually erode away because of human caused global change (Fig. 5.8b).

The same history is true for all of the main river deltas in the Mediterranean Sea, because of human dominance over the normal geologic processes [26]. Human civilization that has existed for thousands of years around the Mediterranean has changed the vegetation of all the main river drainages including the Ebro of Spain, Rhone of France, Po of Italy, and Nile of Africa [24]. The most important of these human changes has been the building of dams. As a result, the long-term cycle of delta building from sediment input of the rivers is disrupted. Because of dam building along all of these rivers, the sediment from the rivers is trapped behind the dams and is not carried to and deposited in the deltas. For example, as mentioned before, the annual Nile River floods previously deposited a large yearly amount of sediment on the delta prior to the Aswan Dam. This yearly deposit caused the Delta to grow into the sea and produce rich farmland, not only in the Nile Delta, but also for the other main deltas of the Mediterranean Sea, the Ebro, Rhône and Po river deltas.

As a result, all of these deltas are now gradually eroding away because of the lack of sediment input. The most severe problem is for the Nile Delta, which is predicted to have the delta edge retreating as much as 80 km from its present position in the next 50 years [27]. This results from a combination of human caused and natural processes. When there are no dams, the input of sediment to the delta overcomes the natural process for the delta to sink because of the weight of sediment on the Earth's crust. When this natural input of river sediment to the delta is trapped behind the dam, the delta will sink and the shoreline will recede. Now with global warming, sea level is gradually rising because of the melting glaciers and the warming and expanding sea, which also causes the shoreline to recede. In sum, as all of these deltas recede because of human-caused global change, the richest farmland of the Mediterranean Sea coastal areas will gradually be lost.

As described in the summary of global warming (Chap. 5) there also is a significant human caused loss of farmland in the northern African region adjacent to the Mediterranean Sea. This loss results from desertification in the Sahara region, because of the severe droughts related to global warming, as well as poor agricultural practices. Consequently, the strong desert winds erode and transport huge quantities of dust from the Sahara northward into southern Spain. Of course we are breathing this dust-laden air and new studies show that hospital admissions increase during and after dust storms [28]. Equally bad, several times a year the air at our home in Granada Spain and elsewhere in southern Europe sometimes turns reddish brown and then it rains mud so that you cannot see out of car windows. Everything is covered in this fine red mud and sometimes it takes a power washer to clean it off. During the past 15 years that I have lived in southern Spain, this has gotten worse every year. This is yet another example in my lifetime that global warming is already significantly affecting our Earth's environment.

Another global change, that has taken place during my four decades of experience with the Mediterranean coast of Spain, is that there has been an increase in the population of jellyfish. When I lived in Spain in 1981, we often went to the small coastal town of Cadaqués in Spain just south of the southern French border, which also was the home of the famous painter Dalí. In 1981, the summer vacationers all swam happily in the sea. The last time I visited Cadaqués in 2005, I observed that swimmers headed happily into the sea and then immediately left screaming. When I went to the beach I saw that there were thousands of jellyfish covering the beach. Thus when swimmers went into the sea, they were immediately getting stung by masses of jellyfish and left screaming. This increase of jellyfish in the ocean has been observed worldwide and attributed to global warming of the sea, acidification, as well as human contamination from fertilizers and sewage. In the case of the Mediterranean Sea these amounts of jellyfish have not been observed in the last two centuries. (www.theguardian.com/environment/2015/aug/21). The jellyfish are less affected by the global change factors whereas their natural predators are. These global changes result in an increase of less desirable species that upset the normal ecology of the ocean.

I had another experience with human-caused pollution in southern Spain, when much to the consternation of my new wife I spent part of our honeymoon sampling the river systems that drain the Rio Tinto mining region. The Rio Tinto and Rio Odiel rivers travel 50 km (30 miles) to the Gulf of Cadiz in the Atlantic Ocean [75]. Because of my experience studying toxic heavy metals (e.g. mercury) in Alaska, I was curious what the contamination differences were between the natural erosion of sulfide mineral deposits (e.g. arsenic, copper, lead, zinc), compared with those caused by mining. This was an ideal place to make this comparison, because the Rio Tinto mine is the world's longest continuously operating mine for over 5000 years. The mine also is in the world's largest sulfide deposit and it contains the toxic elements of lead and arsenic.

The Phoenicians began mining in the Rio Tinto and gave one of the rivers draining this area the name of Rio Tinto, because the river is the color of red wine, which is called tinto in Spanish. We sampled the river sand to see the effect of the naturally eroding sulfide minerals and how far downstream from the deposit high levels of these toxic elements could be found. We sampled the riverbank mud to see how the river transported metals in the fine clay that was suspended in the river water. It was in the middle of summer when the temperatures were 45 °C or well over 100 °F. At these extreme temperatures we hardly saw anybody whom was crazy enough to be outside and we each were drinking a liter of water every hour.

We sampled along the entire lengths of the Rio Tinto and Odiel rivers and then down into their estuaries and along the beaches near the river mouths. I found that like in Alaska, the sand mineral grains did get diluted downstream as other sediment was washed into the river [29, 75], . However, the fine mud contained amounts up to several thousand parts per million of lead and arsenic. Consequently, the river valleys were highly contaminated and I have always wondered what was happening with people living in the valleys and eating the goats and sheep that grazed in the toxic valleys.

I did find that by the time the beaches were reached, the heavy metal pollution was no longer detectable. I wrote a paper describing my results and a USGS postdoctoral fellow reviewed it. He had done his PhD thesis looking at the chemistry of the inflowing Atlantic water that passed my Rio Tinto study area and entered the Gulf of Cadiz. His thesis and published articles tried to explain the sudden influx of the toxic elements into the Gulf of Cadiz, but he did not know about the history of the Rio Tinto mines and rivers, which had introduced the toxic elements. After reviewing my paper, during his first opportunity of Christmas vacation, he rushed and hired a boat to take samples and provide the correct explanation for what caused this anomaly in his thesis data. This is an example of how widespread global knowledge is necessary to explain global changes.

Because of my new association with the USGS post-doctoral fellow, we decided to further investigate the history of the contamination from the Rio Tinto mining district. Geologists have estimated that the original sulfide deposit was approximately 750,000 tons. However by the time the Phoenicians and Romans discovered it, only about 250,000 tons remained and the other two thirds of the sulfide deposit had been eroded into the sea [75]. As part of the USGS/Spanish project, we had taken cores from the surface mud layer on the continental shelf in the Gulf of Cadiz. I thought that because 500,000 tons of the Rio Tinto sulfides had been eroded and deposited into this mud, it should be contaminated from this natural erosion.

To our surprise, we found that most of the deeper mud showed no contamination and that the pollution only began close to the seafloor surface mud, which had an age in the late 1800s [30]. We then looked at the Rio Tinto mining records and found that the increase in mining began in the late 1800s and paralleled the increase in pollution. So what we discovered was that we could show the beginning of the Industrial Revolution and its affect on the offshore pollution of mud. There was no affect of pollution from natural erosion or the mining of the Phoenicians and the Romans. There had been significant mining by the Romans during the time when Caesar controlled the Iberian Peninsula. The fortune he obtained from mining in the Rio Tinto area allowed him to fund his armies and return to conquer Rome. So the Rio Tinto mining area not only had interesting implications for human-caused global change of the industrial revolution, but also had a significant impact on Roman history.

The aforementioned study of the Rio Tinto contamination was an offshoot of the second part of the Spanish project, which studied the Gulf of Cadiz area in the Atlantic Ocean off Spain [31]. Again the project began with the rivers at the coast and extended into the deep sea. A large part of the project focused on the Mediterranean Outflow Water (MOW) that exits from the Strait of Gibraltar and then travels along the Iberian continental slope northwest towards Portugal. The MOW is important for the global water circulation of the Mediterranean Sea and Atlantic Ocean [32, 39]. Studying the history of the MOW is relevant for global change because the MOW flow is controlled by cold and warm climate change of the glacial and inter-glacial ages and this climate control has future implications related to global warming.

This important water circulation system begins with the cooler, lower salt content and lower density Atlantic water traveling eastward along the Gulf of Cadiz continental shelf of Spain [32, 39]. The Atlantic water enters the Strait of Gibraltar at the top and continues to circulate eastward to the end of the Mediterranean Sea. The Mediterranean climate results in high evaporation, which leaves the salt behind and makes the MOW water much denser in the eastern Mediterranean Sea. This water then sinks and travels along the bottom of the Mediterranean back to and through the bottom of the Strait of Gibraltar. As a result, if you take a profile from the top of the water to the bottom at the Strait of Gibraltar, you have a great difference in the density between the top Atlantic water and the MOW water at the bottom of the Strait.

The Germans during World War II were very clever and had their submarines travel near the bottom through the denser MOW when they entered into the Mediterranean Sea through the Strait of Gibraltar. The allies never understood how the German submarines could get into the Mediterranean Sea without being detected by sonar. However this strong density contrast of the deeper MOW did not allow the primitive sonar systems to detect the German submarines. In other words, the strong density contrast caused the Allied sonar pings to bounce off this denser MOW boundary and the German submarines could not be detected.

Similar to the Bering Strait, the large amount of water squeezing through the Strait of Gibraltar results in strong currents that sweep all of the sediment out of the Strait so that only bedrock is found. In 1990 we studied this bedrock surface because there were some proposed plans to put a bridge across the Strait of Gibraltar. This study was important because it showed that the strong currents would probably cause too many environmental problems for a bridge. This is an example of the importance of environmental assessment prior to infrastructure projects so that problems of human-caused global change are avoided.

One important aspect of the MOW history that we discovered also was related to the Strait of Gibraltar. We found that there was a major change in the MOW system as sea level lowered about 120 m during the Pleistocene ice ages. The sea level lowered because ocean water was transferred into the large continental glaciers more than 10,000 years ago. When sea level was lower in the shallow Strait of Gibraltar, the circulation through the Gibraltar Strait was more restricted and the strong MOW flow did not develop [32, 39]. As a result, mud was deposited on the continental slope instead of the MOW sand layers that we found at the present seafloor surface. We could look at the subsurface below the seafloor in our seismic profiles and see the evidence for the different ice ages that took place during the past 2 million years. We could see a set of sand waves developed during a high sea level with strong MOW currents, and then a mud layer deposited during an ice age of low sea level and weak MOW currents. We saw four cycles of sand waves developed during periods between the glaciers. In this way we could find out important affects of climate history on the MOW deposits and circulation of the Atlantic Ocean. The MOW in the future may be affected by both the rising sea levels and increased evaporation from global warming.

4.7   Gulf of Mexico

In 1990 I began research in the Gulf of Mexico (GOM), which has continued until the present. These investigations had three main phases. The first in 1990 with the USGS group from Woods Hole, Massachusetts, was to study the submarine Mississippi Fan on the deep-sea floor of the northern GOM (Fig. 4.14). The second phase with the same USGS group in 1997 was to look at the small confined basins on the continental slope that are called minibasins (Fig. 4.15). The third phase of research has been to summarize the

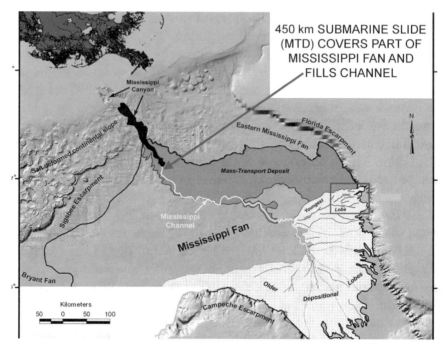

Fig. 4.14 Map of Mississippi River Delta, Canyon, Channel and Fan, which shows the large mass transport deposit (MTD in orange color) from a massive submarine landslide that covers most of the youngest channel and fan with a sheet of debris. The yellow region is the youngest distal lobe at the end of the Mississippi Channel and the red box outlines channel splays at the end of lobe. Figure source is Twichell et al. [66]

deepwater deposits of mini-basins and submarine fans in the northern GOM (e.g. [33]). These studies were conducted with colleagues at the University of Texas at Austin and the University of Texas at Arlington.

The first two programs studied the deep-sea sedimentary systems related to turbidity currents that generate in submarine canyons, travel down the canyons into channels on the deep seafloor, and then deposit sand layers in submarine fans, or mini-basins (Fig. 2.7a) [34]. These types of depositional systems are called turbidite systems. During the past few decades, turbidite systems have become important for petroleum resources, especially in the northern GOM, Brazil, West Africa, and Indonesia. Turbidite systems are good producers of petroleum because they combine sand layers of turbidites with organic-rich mud layers in between (Figs. 2.6c, 2.7b) [34]. These are the first two important elements for a petroleum system, because the organic rich marine mud produces the petroleum source material and then the petroleum migrates into the sand layers that become the petroleum reservoirs. The third

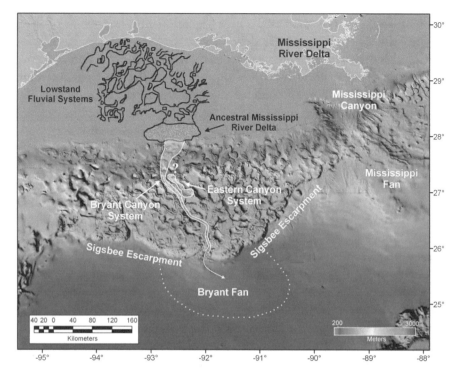

Fig. 4.15 Three dimensional rendition of swath bathemetry shows the Louisiana coast, Bryant Fan and Mississippi River Delta feeding the present day Mississippi Canyon and Mississippi Fan in the northern Gulf of Mexico. To the west is the ancestral delta and lowstand fluvial system of the Mississippi River that formed during the previous ice age and low sea level from 130,000 to 160,000 years before present. At this time the ancestral Mississippi fed the Bryant Canyon and East Canyon chain of mini-basins and eventually the Bryant Fan in the deep Gulf of Mexico. The multiple depressions on the continental slope to the north of the Sigsbee Escarpment are mini-basins. Figure source is Nelson et al. [65] modified from Liu and Bryant [69] and Suter and Berryhill [70]

element of a petroleum system is some kind of geometry or structural trap, such as uplifted dome structures, where the oil is trapped in the turbidite sand reservoirs.

For the study in 1990 we used sidescan sonar towed near the sea bottom that provided detailed information on the channels of Mississippi Fan (Fig. 2.5) [35]. Bottom transponders, that emitted location signals, were put on the seafloor. These signals provided accurate navigation to within a few meters. This allowed us to go back to specific sites we had observed in the sidescan and then obtain cores of sediment at these locations. First, we studied where sheets of debris from gigantic submarine landslides in Mississippi Canyon that had traveled down and covered much of Mississippi Fan, including some of the channels. These gigantic slides happened occasionally on Mississippi Fan and the debris sheets covered areas as large as one third of the 600 km diameter of the Fan (Fig. 4.14). The debris traveled into the deep sea for distances up to 450 km down the fan from the mouth of Mississippi Canyon at the base of the continental slope. Although these events occur hundreds to thousands of years apart, they have the potential to create gigantic tsunamis in the northern GOM. The second location we studied was the very distal fringe of Mississippi Fan, where sand still traveled down small channels with only a few meters of relief (Fig. 4.14, see red box). We were surprised that submarine landslides could create debris flows that followed these channels and traveled to the end of the fan 600 km from the base of the continental slope [36].

Our discoveries have important implications for ancient submarine fan petroleum reservoirs. From the less detailed seismic profiles of petroleum companies, it appears that the distal regions of submarine fans contain large sheets of sand. However, our studies indicate, that the apparent sheets of sand, are complex with multiple small channels and thin layers of muddy debris flows (Fig. 4.14, see red box) [36]. The oil companies expect that these apparent sand sheets provide large continuous reservoirs, however our studies show that the distal submarine fan reservoirs can be complex. Our results have become important for large new GOM oil discoveries that have been made recently in old ancient submarine fan systems that lie more than 30,000 feet below the seafloor.

Our GOM cruise in 1997 concentrated on the small mini-basins of the continental slope along the north central GOM (Fig. 4.15). For these studies, we took high-resolution seismic profiles across 20 mini-basins and then took long cores at selected locations in these mini-basins (Figs. 2.5, 2.6b, c) [37]. For the past several decades, there has been a huge amount of petroleum exploration and development in the GOM mini-basins. The oil companies have had great success with some of the largest USA petroleum discoveries during the past few decades in the GOM mini-basins. However, the company's lower resolution seismic systems cannot provide the necessary details for development of the reservoirs in these mini-basins. Our studies provided much more detailed information on the sand reservoirs that pond in these basins and the channels and canyons that feed these mini-basins.

4.8   Gulf of Mexico Hazards

For the 1997 research cruise, we left from Galveston at night and it looked like we were traveling through New York City with the entire offshore area lit up with the thousands of petroleum platforms. These platforms extend for tens of kilometers offshore, although as you get into deeper water there are fewer platforms. One night when we were far from shore with no platforms in sight, we saw a huge red glow in the sky, which the captain thought was a fire on one of the platforms. The rule of the sea is you always have a first priority to rescue anybody in danger. Thus the captain turned away from our work and headed towards the huge orange glow. Research ships travel slowly at only 16 to 24 km (10 to 15 miles) an hour, and thus we traveled for four or five hours until we finally could see that the petroleum platform was just flaring or burning off methane gas in a normal manner. It may not have been normal because perhaps they were doing this at night when nobody was nearby to see that they were flaring so much gas. I was unhappy to see this example of gas flaring, which is extremely wasteful and even worse, an unnecessary contributor to global warming.

Another adventure took place at the end of our 1997 cruise when we were traveling the 100 km (62 miles) back to port. A hurricane was starting so that we needed to get to port as soon as possible. However suddenly in the middle of the night, the ship went dead with no engines operating. As I mentioned earlier, whenever a ship goes silent like this you sit bolt upright in bed in fear. Especially in this case if we had no engines in a hurricane, it probably would have been the end of us. Luckily the ship had two engines and eventually one engine began working, but of course we could only go at half speed towards port to escape the hurricane. We eventually reached the port, but by this time we had hurricane force winds and only one engine to operate the ship. As a result, even though we were only a few tens of meters from the dock, because of the strong winds, it took two hours to get the ship alongside the dock. We got off the ship immediately because some of us had flights from the Houston airport. However, we had to drive through the hurricane rain and by this time some of the freeways began flooding, but we managed to drive through the water and get to the airport. This did not matter much because all the flights were canceled, so we ended up sleeping in the airport anyway.

Because of these increasingly severe hurricanes in the northern GOM from global warming, there are many global change problems related to the thousands of offshore petroleum drilling platforms and thousands of kilometers of pipelines. This infrastructure is subject to geologic hazards from the severe hurricanes that occur in this area. The increase in category 4 and 5 hurricanes and related hazards has caused loss of life and significant damage to the offshore infrastructure of the petroleum companies.

The geologic hazards result mainly from the unstable muddy sediment deposited in the northern GOM. This sediment is subject to submarine landslides, even on gentle slopes of the continental shelves. In addition, this organic-rich mud can become charged with methane gas, which makes the mud even less stable. When the high waves of the strongest hurricanes pound on the ocean seafloor, it sets off submarine landslides. When these landslides occur, they can result in failures of drilling platform legs and can break pipelines (Chaytor 2020). Consequently, with the onset of more intense hurricanes like Katrina in 2005, this hurricane alone destroyed or damaged up to 30% of the offshore petroleum infrastructure including 109 oil platforms and five offshore drilling rigs and over half of GOM production was shut down (www.offshore-technology.com/features/offshore-oil-recovering-harvey/). Because damage to pipelines and resultant oil spills do not have to be reported, there is no way to monitor environmental damage when these spills are far from shore.

These aforementioned data show the importance of regulations so that hurricane damage and oil spills can be assessed. Also better regulations can help prevent large oil spills and provide information collected from previous oil spills. An example is the British Petroleum Company (BP) Deepwater Horizon oil spill in 2010, which resulted in the largest oil spill in history. As is common with these oil spill events, a number of factors combined to cause the BP spill. BP was in a rush to complete this horizon well, even though it was known that very high sub-bottom pressures were present when drilling the well. There are a number of ways to deal with these high pressures when completing or capping the well after the drilling is finished. Cement is poured down the well to plug it, and there is a blowout preventer structure that is at the top of the wellhead in case the cement plug fails.

Even with these backup systems for the BP Horizon well, both systems failed and the well blew out, which resulted in oil spilling out for weeks. The spill may have been prevented if there were better regulations. The regulation at the time of the BP spill was that there should be a test of the blowout preventor at least two weeks prior to capping the well. It would seem prudent that there should be a regulation that there is a test of the blow out preventer immediately before a company intends to cap a well. The second contributing factor to the blowout and oil spill was that the cement was substandard and had failed tests (https://en.wikipedia.org/wiki/Deepwater_Horizon_oil_spill). In the rush to complete the well, BP used the substandard cement. It seems there should have been a regulation in place to provide data on the cement tests and prevent the use of substandard cement if it failed tests.

Another factor that made the oil spill worse was that the EPA originally permitted BP to use chemicals to disperse the oil at the surface of the ocean. Of course BP used dispersants, because if the oil spreading over thousands of square kilometers were dispersed, the spill would be less visible and would seem less severe. I was appalled to read that they were using these dispersants, because I remembered that dispersants were used after the first large oil tanker spill of the Amoco Cadiz, which occurred several decades earlier. The scientific reports on the Amoco Cadiz spill found that the chemical dispersants caused more damage than the oil spill.

Eventually, the EPA rescinded their permission to use the chemical dispersants when they realized the environmental damage it was causing. When the petroleum is dispersed into small molecules, the molecules can be incorporated by microscopic animals living in the spill area and incur damage or death. When the spill is not dispersed, the larger animals can tend to avoid the spill. It is sad that previous scientific knowledge is not utilized and regulations put in place to prevent the recurrence of preventable environmental problems. For scientists dealing with factual knowledge, it is incomprehensible that the populist 2017 to 2020 USA government prefered to eliminate government environmental regulations at the risk of the Earth's living organisms, including humans.

Another example of the use of scientific knowledge to prevent more damage from major oil spills is provided by the Exxon Valdez oil tanker spill in Prince William Sound in southeast Alaska in 1989 (Fig. 3.2). Again, this was one of the largest oil spills prior to the BP Horizon spill in the Gulf of Mexico. One of the results was the severe oil coating of the coastal rocky cliffs. In these cliff areas, steam cleaning was used to take off the oil, because it was impossible to mop oil off the cliffs in the same way that they did on the sandy beaches. A decade after the Valdez spill, I attended a meeting on the environmental effects of the spill. At the meeting it was reported that steam cleaning sterilized the cliffs and there was no recovery of the biota after 10 years. In contrast, in areas where oil had not been cleaned off the cliffs, there had been a full recovery of the biota, because the strong waves of Southeast Alaska had naturally cleaned off the oil. This provides yet another example for the reasons to have thorough studies for environmental problems. Unfortunately, the continuing budget cuts in the USA federal government are reducing these necessary environmental studies.

4.9   Lake Baikal Russia

I spent the summers of 1991 and 1992 undertaking research in Lake Baikal, Russia. This was a cooperative USGS project with the Russian Institute of Limnology in Irkutsk to determine the climate history in the interior of the Asian continent. Lake Baikal is found in Central Asia near the boundary of Siberia with Mongolia (Fig. 4.16). It is an ideal place to determine long-term climate history in the center of a continent, because the lake is the world's oldest (>20 million years) and deepest (up to 1700 m or a mile deep). It is 636 km (395 miles) long and 79 km (49 miles) wide and contains 20% of the world's fresh water because of its great depth and volume (https://en.wik ipedia.org/wiki/Lake_Baikal).

Other lakes like the Great Lakes in North America have much larger surface areas, but are only a few hundred meters deep. Lake Baikal is situated in a rift zone, similar to the setting of the African rift lakes. These geological settings are locations where continents are ripping apart. When this happens, the rupture results in deep basins, which in the case of Africa and Baikal, then become occupied by deep lakes. Our USGS study in Lake Baikal obtained high-resolution seismic profiles and took cores of the sediment of the deep lake floor (Figs. 2.5, 2.6b, c).

The deep lake bottom deposits are some of the world's best examples of the basic types of turbidite systems and Lake Baikal is an ideal laboratory to study these. Because Earth's crust is ripping apart, the deep rifts like Baikal have large faults on each side. On one side of the lake there are border faults that have steep slopes averaging 30° and extending with a total relief in the case of Baikal to over 2000 m (6,500 feet) above water and up to another 1700 m (1 mile) below water (Fig. 4.16 see no shallow water on western side). The other side of the lake is a more gradual ramp and has many local rivers entering the lake as well as the large Selenga River that drains parts of Mongolia (Fig. 4.16).

The reason Lake Baikal is so ideal to study turbidite systems is that you can look up at the mountains on the western border fault and see where massive landslides have come down the steep slope to the lake (Fig. 4.17). Then in the seismic profiles you can see small apron deposits of a few kilo-meters diameter on the deep lake floor that are deposited by slides below water [38, 32, 39]. Thus you can see the entire sediment system from the mountain source to the apron deposits of the deep lake floor. These aprons of sediment on the lake floor are made up of sheets of sand and gravel like you see in Crater Lake (Figs. 4.2, 4.4) and similar to those of alluvial fans at the base of high mountains on land. The reason that these sediment aprons at the

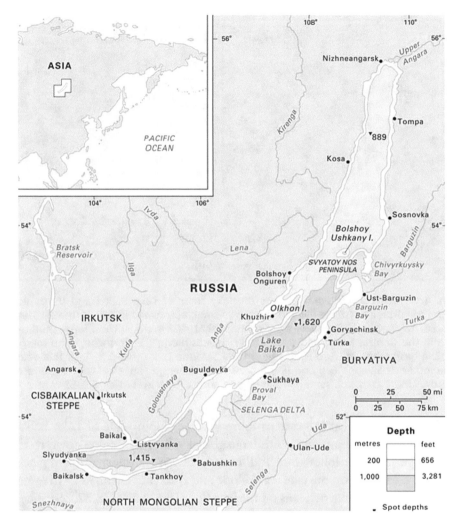

Fig. 4.16 Map shows the location of Lake Baikal in the Russian area of Siberia in central Asia (see inset map). The lake is divided into three basins, Southern, Central, and Northern by ridges called accommodation zones (see Fig. 4.17). The Southern Basin is located to the south of Selenga Delta and the Central Basin, which is the deepest (1620 m, 5,265 feet) is located to the north of Selenga Delta. The North Basin, located north of Bolshoy Island, is the shallowest (889 m, 2,890 feet). Figure source is Britannica.com

base of the lake slope are important is that they can provide ideal petroleum reservoirs in the right setting where petroleum can collect. Aprons provide important petroleum reservoirs in the North Sea, off Brazil and in the GOM. Consequently, studying the details of these aprons is relevant for petroleum resources.

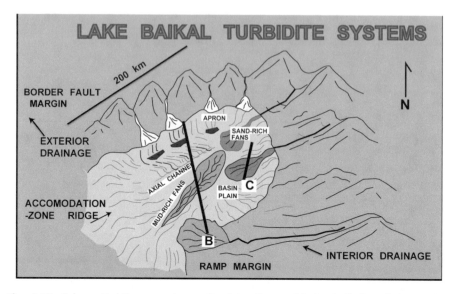

Fig. 4.17 Schematic diagram shows the lake floor of Lake Baikal and the typical types of turbidite systems located there (see small aprons along the northwest steep border fault side of the lake, axial channel that drains the aprons, large mud-rich fan in the Central Basin, and small sand-rich fans along the northeast ramp margin of the basin). The letter B is located on the Selenga Delta (see Fig. 4.16) that feeds the mud-rich Selenga Fan and line B crosses the Selenga Fan, axial channel, and an apron. Line C crosses a sand-rich fan. Figure source is Nelson et al. [32, 39]

When you look at the eastern more gradual sloping side of Lake Baikal, you can see high mountains with large glacial valleys that come down to the lake edge. You can follow how sediment has been transported down glacial valleys to canyons on the lake basin slope and then into channels of the small 20 km diameter sand-rich fans on the lake floor (Fig. 4.17) [38,32, 39]. These sand-rich fans provide some of the best petroleum reservoirs in the world such as onshore and offshore in Southern California, the North Sea, and Brazil.

On the same gradual east sloping side of Lake Baikal, the Selenga River deposits a 20 km (12 miles) wide delta (Fig. 4.16). Sediment from this large Siberian river is transported down the lake slope in canyons and then into channels of the Selenga Fan on the deep lake floor in the Central Basin north of the Selinga Delta (Fig. 16) [38, 32, 39]. The Fan is 80 km (50 miles) long and mud-rich, because the Selenga River transports a larger amount and finer sediment compared to the other smaller rivers feeding sand-rich fans in Lake Baikal. As a result, Lake Baikal contains examples of the three basic types of turbidite systems, aprons, sand-rich fans and mud-rich fans (Fig. 4.17). These systems can all be compared in one basin and provide examples for subsurface ancient petroleum reservoirs in similar rift basin settings.

In all the types of turbidite systems of Lake Baikal you can see the same climatic control, because the main deposition took place during Pleistocene glacial times of colder climates compared to the warmer Holocene interglacial climate of the past 10,000 years when little sediment is deposited in these systems [38]. This cold Siberian climate control of Lake Baikal deposition contrasts with that in the warm tropical setting of the African rift lakes. In the African lakes the change of glacial climates resulted in the lakes nearly drying up and having only shallow water depths. However, during the warm tropical climate of the past 10,000 years, the rift lakes have filled up with water to become deep lakes with turbidite systems. This contrasts with Lake Baikal in which the turbidite systems do not have significant deposits during the warm periods.

The deposits in these two different rift lake settings show the complexities of natural climate variability because turbidite deposition diminished in the polar Lake Baikal and began in the tropical African lakes during the same time of warm climate change. Such differences can be confusing for some people to understand and lead them to deny human-caused global warming if they do not look at different multiple worldwide patterns. For example, different patterns of climate change can vary from one location to another at the same time, such as an extreme El Niño causing torrential rain in one location or drought in another, or a polar vortex breakout causing bitter cold in North America and unusual warming in Europe and Asia (see Chap. 5 Sect. 5.7 for further discussion).

4.10   Russian Adventures

There were many adventures related to our work in Russia. My first impression with this first visit to Russia was to notice all the crumbling bridges and buildings, and the poor state of infrastructure. This made me wonder why did the USA spend all this money for the Cold War to combat a country that seemed to have such decaying infrastructure. My next impression was at the Russian Academy Nauk hotel in Moscow after flying continuously for almost two days to get there. This Russian government hotel was where scientists stay when doing cooperative work with Russian government scientists. As with typical transatlantic flights, I ended up with a terrible migraine headache, so I went to bed immediately to get a few hours sleep because we had to go out to the airport at three AM to get our flight to Siberia. After sleeping a few hours I awoke sick from the migraine headache and turned on the light. The bed and floor were a brown sea of cockroaches, which did not make me feel any better. In addition, I had many bites on my skin caused by the cockroaches.

After I got up, we took a wild ride to the Moscow domestic airport in a taxi. I say wild ride because whenever the driver encountered a car in front of him as he sped down the road, he just passed whether there were oncoming cars or not. As a result, on this two-lane road, at least three cars were forced into the ditch on the ride to the airport. The airport was a scene of chaos with hundreds of people and families camped out on the floor. It soon became apparent why there was this chaos, because passengers had to wait until a plane was full and then the plane would leave. Schedules did not seem to mean anything nor did previous reservations. It took hours and one by one to confirm our reservations as long as we continued to bribe someone.

The second year when flying in Russia, we again could not get on the plane, but we had a resourceful USGS cook with us. She had no end of things to bribe airline officials with, such as nylons, cigarettes, money, and jewelry, until finally all of us got on the plane. Once on the plane to head across Siberia through five time zones, things were not much better. The planes were in terrible shape with oxygen masks dangling from the ceiling and when we took off, the seat of one of my USGS colleagues broke off and he flipped over backwards. When we landed halfway across Siberia to get more fuel, the ground crew rushed out and put fire extinguishers on the brakes. The food, which we called rubber chicken was no better and on all the flights never varied. It was interesting that a few years after our flight on Aeroflot, the US State Department banned all USA government employees from flying on Aeroflot because of the poor safety record. Fortunately now, Aeroflot seems to have improved.

Once we arrived at the ship on Lake Baikal, our first task was to install water filters and toilets in the bathroom, because the sanitary conditions were bad. The previous year people had gotten sick with dysentery and one scientist was so ill that they took her to the hospital. They had to try several different hospitals because at each of the regular hospitals, the doctors immediately wanted to perform an appendectomy, which would have killed the person. Finally, my good Russian friend Dr. Eugene Karabanov, who we called Genya, took the scientist to a children's hospital where they provided the proper treatment of rehydration and thankfully the person recovered.

The conditions on the Russian ship were as terrible as the domestic airport in Moscow. It appeared that the ship had never been cleaned. And also during and after the cruise, we scientists suffered skin rashes, scabies, crab lice and other maladies. However the worst problem was that everybody in the scientific party at one time or another became ill during our research cruise, except for our chemist and myself. He packed Evian water with his equipment and I followed a suggested protocol to disinfect my water with iodine.

In fact these illnesses were to continue for nearly a decade after our cruises, until they found that the ship water supply was being contaminated with ship sewage. There was a horrible smell of sewage throughout the ship from the poorly maintained bathrooms. In fact the only time a bathroom smelled good was one time when the alcoholic Russian cook in desperation drank her perfume for alcohol and then was sick in the bathroom. In a previous year, some of the Russian crew drank denatured alcohol and nearly got poisoned to death. I thought it was a telling sign when I observed that the Russian vodka bottles have no way to be closed again once they were opened. There was no screw top or cork to close them, because obviously once any vodka bottle was opened, it would be finished.

The alcoholism of the Russians caused problems throughout our cruise. Whenever we were in port, most of the crew would get completely drunk. In one instance, while we were in port and the ships chief engineer was gone, we had to weld our large air compressor to the ship deck for our seismic profiling. When the chief engineer came back to the ship completely drunk, he took a sledgehammer and began smashing our air compressor equipment because he didn't like the location where we had welded it on the deck. Then he proceeded to fall 10 m (30 feet) overboard into the lake and we had to rescue him and put him back on the ship to sober up. The captain unlike the other crew was responsible, however the first night when we left to begin our work, he suddenly had to leave for an emergency. This left the drunken first mate to run the ship and around midnight I noticed the ship was zigzagging down the lake. I went to the bridge and found the first mate drunk on the floor, so that no one was steering the ship.

The other interesting thing that I learned a year later when talking to my Russian friend Genya, was that the fishing stop we made each day had another purpose rather than feeding the poor ship crew families. The fish that the ship crew caught were not for the families, because any time we came ashore at a village, the crew traded fish for vodka. However, for the scientists these fishing stops were a great break, because while the crew fished for several hours, we could go ashore and hike in the beautiful pristine forests around Lake Baikal. Also the fishing methods were unique because they would run the large research ship up onto the rocky beach and then kept the prop slowly turning so it created a current like a stream coming into the lake, which attracted the fish to catch.

Another unusual experience for a scientific ship cruise, like taking a hike on land each day, was that a small boat would bring out fresh bread. One day we saw the bread boat approaching, but instead of seeing the captain driving the boat, we saw that the cook was driving the boat. When the boat arrived we saw the same old problem, the captain was drunk on the floor of the boat, but fortunately the cook could drive the boat to us. Another amusing incident was another day when the bread boat approached our ship and a young woman was waving frantically. It turned out that one of our electronics technicians had met her the year before and now she had chased him all the way across Russia and was arriving on the bread boat. The technician was terrified and hid in the ship because he did not want to be found.

Similar to the problems with the contaminated water, the food situation was not much better. In 1990 the first USGS scientific cruise, there was almost no food to eat. At this time just before the collapse of the Soviet Union you would go into a store and the shelves would be completely empty and that is why they had almost no food on the ship. Because of this lack of food in 1990, the Russians tried to provide more food for the cruise in 1991. They brought an entire side of beef aboard the ship and tied an old refrigerator on the deck to hold the beef. Unfortunately, this beef was used the whole month-long cruise and we constantly made comments about having green meat for dinner. As a result, in 1992, the USGS brought a container full of food and our own cook because of the experiences in 1990 and 1991. So we ate well in 1992 and there was an abundance of food and other goods in the stores. Unfortunately, the Russian people had no money to buy food in the stores. Consequently, you would see people out on the streets selling anything that they could, such as paintings and antique religious icons for a few tens of dollars.

Even though the Russians were suffering, they continued to be extremely generous. When we were on shore, they would invite us to their homes and provide good meals from the gardens that they were growing. My friend Genya had collections of coins and Russian religious icons. At the end of the 1992 cruise, he gave me a gift of several icons from the seventeenth, eighteenth and nineteenth centuries. I tried to refuse, but Genya secretly put them in my suitcase. When I was leaving the international airport in Moscow, you had to put your suitcases through a scanner. The Russian government scanned all exiting bags because they were concerned about all of the artifacts like icons that were being sold in desperation by the people and then taken out of the country.

When the security people put my suitcase through the scanner, of course they found the icons and confiscated them. I thought that I may be going to jail, but as always there was another Russian scientist accompanying me and he took back the icons and I was able to get on the plane. However, months later I got a phone call in my office in the United States when the USGS technicians in California were unpacking the scientific equipment from our Russian cruise. They had found a series of icons that they thought belonged to me. My generous friend Genya had secretly put the icons back in our scientific equipment so that I could have them.

Genya was very lucky to have been alive for our cruise in 1991 and later on a scientific cruise in 2002 (Fig. 2.7b). As we were traveling along the lake one day on the ship, he said this is where his research truck fell through the ice the winter before. Normally the ice is two meters thick in Lake Baikal. Driving trucks across the lake ice is the main way that supplies are taken to villages around the lake, because there are few roads on land. Also similar to our Minnesota lake studies, it was much cheaper for the Russians to go on the ice in the winter and take samples. This was what Genya and his students were doing in the winter before our 1991 cruise. Their truck fell through the Lake Baikal ice because year by year the ice has been becoming thinner than previously, which is yet another example of the rapid loss of ice volume that is taking place from global warming, particularly in the Arctic (e.g. Figs. 4.6, 4.10).

The truck was supposed to float, but it did not and Genya and his scientific party were left in the freezing water in the middle of Siberia in the winter. Genya told me that they had a difficult time trying to get out on ice and had to finally slither like snakes across the thin ice to get out of the water. However, this was only the beginning of their adventure, because it was remote and the scientific party had to walk 15 km (9 miles) to the nearest village. By that time they were nearly frozen like icicles, but they all did survive. It showed me again how strong and enduring the Russians are. Unfortunately, my Russian bear friend Genya was not strong enough to survive brain cancer and died in 2012. This was one of the saddest moments in my life when this good friend passed away. We had spent months together on ships and working together in Russia and in the United States and staying in each other's homes.

There were several more adventures at the end of our cruise in 1991. When we had to take all of the equipment off the ship, our air compressor was not on the dockside of the ship. The ship had to be turned around so the side with the compressor was along the dock, and then a crane had to lift the compressor off the ship. It took four days until there was a sober crewman and crane operator at the same time, so that we finally could get the compressor off the ship and leave for the United States. Our sediment cores from the lake floor fared even worse. It took them a year to be shipped back to the USGS to be studied in the laboratory at Woods Hole, Massachusetts.

We took two sets of cores at each apron and fan sample site in Lake Bakial in 1991, one core for Russia and one for USA (Fig. 4.17). While the USA cores were taking a year to be shipped, the Russians gave the second set of cores to the Japanese. They immediately began studying the Russian cores and published some results even before our cores had arrived in the United States. Our chief scientist was completely distraught because he had financed the cruise, yet other scientists not involved in the cruise, had published results before the USGS. When we returned for the Lake Baikal cruise in 1992, we could see what had happened. The Russian laboratory had new cars, computers, cameras and equipment given to them by the Japanese in return for using the cores that we gave the Russians. As shown by the history of our cores, this was an extremely stressful time in Russia and everyone was desperate to survive, including the scientific laboratories. In fact, when we were on shore, we deliberately ate in private homes of Russian scientists that had lost their jobs in the collapse of the Soviet Union. We would pay the scientist's families for our food so that they could survive.

The currency situation also was completely chaotic during the collapse of the Soviet Union. When we first arrived in Moscow in 1991, there were only a few free markets where we could shop. At the beginning of the summer, the markets would only accept black-market dollars and not the Russian rubles. It was difficult to buy anything, because we had obtained rubles before we arrived in Russia. Thus before returning to Moscow at the end of the summer, we switched a lot of our rubles back for dollars so that we could buy things at the free markets. Buying many beautiful Russian lacquered boxes obsessed one of our USGS technicians. However, he was frustrated because he only had dollars to pay for the boxes, when only rubles were accepted this time at the market. He kept negotiating with a young Russian man selling boxes at a card table, but the Russian told him he could not sell the boxes for dollars because it was illegal and there were many KGB agents standing around the market. You could tell the KGB because they stood out with their well-dressed clothes.

Finally, my USGS friend convinced the Russian to sell him the boxes as long as my friend would put dollars in the Russian's backpack that was hidden around the corner of a building. My friend came back and picked up his boxes and then started leaving. I was standing next to the table where the Russian had sold my friend the boxes. Suddenly two KGB agents made flying tackles and smashed the poor Russian to the ground and started beating him. Then the KGB agents started questioning and harassing my USGS friend. Because of the KGB actions, the other Russians in the market started screaming at the KGB agents and then they stopped harassing my friend. We immediately left the area and went walking the five kilometers back to our Academy Nauk hotel.

At the beginning we walked through large crowds, but eventually we were walking alone along the street back to the hotel. I looked back and saw the two KGB agents walking immediately behind us and realized this did not look good. Then the KGB agents came up and demanded that we go to a police station with them. I immediately said nyet, nyet, or no we would not go to the police station, because I knew we might never be seen again. The KGB agents kept questioning my friend and asking for all of his papers and he did not have his passport or anything proving whom he was or what he was doing in Russia. Fortunately, I had my passport and it had a visa explaining that we were scientists invited by the Russian Academy of Science. The two KGB agents then began arguing with one insisting that we should go to the police station and the other saying we should not go to the police station because we had the visas from the Russian Academy of Science. Luckily the KGB agent that wanted us to go to the police station lost the argument and we were allowed to keep walking to the hotel. My friend was terrified and went into the hotel and never came out again for several days until when we went to the airport to leave for the USA.

The last thing I did in Russia in 1991 was to stroll around Red Square and look at all of the beautiful tourist things to see. It was astounding a few days after I got back to the USA, to watch on TV when Yeltsin was in Red Square on a tank and the Soviet Union collapsed, and Gorbachev was no longer in power. This was a global political change that had important implications for human-caused global change. After the collapse of the Soviet Union, it became evident how much environmental contamination had occurred in the communist countries. Pollution was rampant, which caused a significant drop of a decade or more in longevity within some of the communist countries (https://www.rferl.org/a/life-expectancy-cis-report/24946030.html). This is an important historical lesson for the populist 2017 to 2020 USA administration to learn, because of its efforts to eliminate the EPA clean air act and regulations on dumping toxic wastes into waterways.

4.11 Corsica and Var France

In 1994 I was invited by IFREMER, the French government marine science institute in Brest France to take part in a cruise studying the submarine Var Fan offshore from Nice France and the western submerged margin off Corsica (Fig. 4.18). Our studies focused on the human caused global change of the

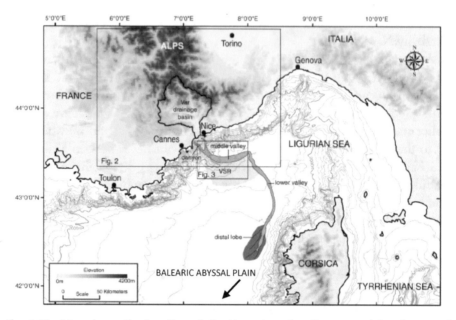

Fig. 4.18 Map shows the location of the Var submarine Canyon and Fan (gray and dark gray pathway) associated with the Var River mouth, which enters the north-western Mediterrean Sea at the city of Nice France (see upper left corner of Fig. 3 box). The French Alps source and Var River drainage basin are outlined (see Fig. 2 box). The Var Fan extends from the Nice source area 200 km (120 miles) south to reach the western side of Corsica (see distal lobe). Note that the right tributary canyon extends into the end of the Nice airport runway where the 1979 subma-rine landslide took place (see upper left corner of Fig. 3 box). The Balearic Abyssal Plain extends several hundred kilometers to the southwest of the Var Fan distal lobe. Figure is modified from Jorry et al. [71]

1979 Nice airport submarine landslide into Var Canyon and the natural caused global change of an ancient landslide off western Corsica.

The Var River drains the French Alps and reaches the ocean at Nice France (Fig. 4.18). There it enters the submarine Var Canyon, which feeds the Var Fan. Our previous 1984 study concentrated on the Var Canyon mouth and the proximal channel of the Var Fan where the effects of the 1979 submarine landslide could be observed (see Sect. 4.6) [23]. My new research with the French focused on the distal end of the Fan where the channel passes into a distal sand lobe that parallels the western coast of Corsica (Fig. 4.18). This new study again used seismic profiling, sidescan sonar, bathymetry and sedi-ment cores to look at the distal Var Channel and the base of slope aprons off Corsica that inter-fingered with the Var Fan lobe (Figs. 2.5, 4.18) [40].

One objective of this study was to follow the history of the 1979 subma-rine landslide and the turbidity current of sand that deposited in the distal channel and lobe of Var Fan near western Corsica (Fig. 4.18). These turbidity currents travel with high velocities of meters (yards) per second and the 1979 and other historic flows have broken telephone cables and disrupted global communications [23]. Most of these currents have been triggered by natural slides, but the 1979 submarine slide apparently had a human cause. This landslide took place in the Var Canyon that is adjacent to the Nice France airport runway. The runway was being extended into the sea to the edge of the canyon and most likely overloaded the canyon slope, which caused the submarine landslide and collapse of the end of the runway [23]. The cause of the landslide was contested in the courts for many years because people died from the tsunami that was caused by the landslide. Consequently, there were no detailed scientific studies for several years.

An interesting side note of the landslide event took place when I had a conversation some years later with a Norwegian geologist friend. He had been on vacation on the Antibes beach across the bay from the end of the Nice airport runway. This was the location where a number people were killed by the tsunami. He saw the water pulling away from the shoreline and realized a tsunami wave was coming and he ran up the hill and escaped. Unfortunately, some other people on the beach did not realize that a tsunami was taking place and lost their lives.

The 1979 event also showed that this modern Var sand-rich submarine fan still has active deposition during high sea level. These sand-rich fans are important analogues for subsurface ancient systems that contain important oil reservoirs at many worldwide locations [34]. The Var Fan has thick sand layers because the Var River drains the French Alps and then transports the sand and gravel directly into the Var Canyon head at the river mouth near Nice (Fig. 4.18). Thus unlike most submarine canyon heads that are not connected to river mouths during the present time of high sea level, the Var Canyon provides a direct conduit to the deep seafloor. Consequently, the Var Fan continues to have present-day sand deposition, such as from the 1979 submarine landslide as well as natural landslides into the Var Canyon.

The historic 1979 submarine landslide provides a valuable analogue where you can trace all the characteristics of fan deposition continuously from the river source to the final depositional sink in the submarine fan. This active fan system thus helps our understanding of ancient submarine fan petroleum reservoirs, which helps maximize the recovery of petroleum. By learning these detailed characteristics from modern analogue submarine fans, during the last 50 years the recovery of petroleum has often doubled from these types of reservoirs. When you double the oil recovery from petroleum reservoirs, you essentially double the world's reserves of petroleum. This is why these detailed studies of modern submarine fans are so important for global energy resources until 100% sustainable energy resources can be developed.

For the same reasons, our study of the base of slope sand and gravel aprons off western Corsica is important, because similar ancient subsurface aprons provide significant petroleum reservoirs, such as in the North Sea. The Corsica aprons also help us to learn about the development of thick massive sand beds that are the best petroleum reservoirs. Massive sand beds form off Corsica because the sand and gravel aprons inter-finger into the thick distal sand lobe deposits of the Var Fan (Fig. 4.18) [40].

The presence of a 30 m thick (60 feet) chaotic layer at the same subsurface depth below the near-surface aprons is another important observation from our seismic profiles along much of the western Corsican margin (Fig. 4.18). This massive layer is interpreted to be chaotic sand and gravel derived from the western margin of Corsica that rises 2500 m (4000 feet) above the shoreline in mountains and this same steep slope continues another 2500 m (4000 feet) below sea level. Consequently, all of the landslides above and below water from this steep scarp shed coarse sediment into the base of slope aprons. This is evident in our cores from the Corsica aprons, which dominantly contain sand and gravel, the same as we interpret made up the massive subsurface bed along Corsica [40].

A few years later, my discovery of the massive bed off western Corsica became significant. An English scientist took seismic profiles and long cores in the Balearic Abyssal Plain and discovered a similar unusually thick sand bed spread hundreds of kilometers (or miles) across the abyssal plain (Fig. 4.18) [41]. He found that this thick sand bed was about 25,000 years old by obtaining radiocarbon ages. He also determined that this layer became thinner and the size of the sand grains in the bed became smaller from north to south across the Balearic Abyssal Plain.

I realized that both of these trends indicated a northern source for the submarine landslides and the consequent turbidity currents that deposited the thick sediment layer over the Balearic Abyssal Plain in the Mediterrean Sea (Fig. 4.18). When I estimated the age of the massive bed off Western Corsica, it was the same age as the thick sand bed of the Balearic Abyssal Plain. The age and the trends leading to the Western Corsica massive bed indicate that the thick sand bed originated from catastrophic landslides off the high mountains and steep slope of western Corsica (Fig. 4.18). The only way that multiple gigantic landslides can trigger synchronously for 100 km (62 miles) along western Corsica is a great earthquake. Because of my interest in determining the history of great earthquakes by examining deep-sea deposits, I suggested that the French study the Corsican mountains to determine if

Corsica could be the source of the widespread massive subsurface bed and the correlative thick sand bed apparently triggered by a catastrophic earthquake about 25,000 years ago.

The study for a Corsican earthquake origin never was done, because of budget cuts and downsizing of the USGS. Such a study would have significant implications for earthquake and tsunami hazards in the Western Mediterranean Sea. Now with the Italians, we have verified in the Eastern Mediterranean Sea that the 365 A.D. Crete earthquake and tsunami resulted in global destruction of the coastal cities there. For example, historic records describe boats on top of houses as part of the total destruction of Alexandria, Egypt [42]. A similar catastrophic tsunami may have occurred in the Western Mediterranean Sea in prehistoric times about 25,000 years ago. Now with the majority of the population living in coastal regions around the Western Mediterranean, the potential earthquake and tsunami hazards should be investigated.

Unfortunately, because science budgets have continued to be reduced in the United States and throughout the developed world, these potential catastrophic hazards have not been fully assessed. Without enough earthquake hazard studies, there has not been enough planning to reduce the loss of life and the risk to infrastructure. The Sumatra earthquake of 2004, that killed 228,000 people is an example of the result of this lack of science funding for studying hazards and planning to reduce the risks (https://en.wikipedia.org/wiki/2004_Indian_Ocean_earthquake_and_tsunami). At a minimum, with a small amount of funding the populations of Asia could have been educated about warnings for a tsunami. People could have been taught, that if the sea suddenly withdraws, they should head immediately to high ground to be saved from drowning. The same education about tsunami hazards should be undertaken for the coastlines around both the western and eastern Mediterrean Sea, because our studies of the Ionian Sea and off Corsica show the potential for catastrophic tsunamis in both areas (Fig. 4.18) [40, 43].

4.12 Antarctic Natural World

In 1994, I was asked if I would like to join Cheeseman Ecotours in Antarctica as a geology lecturer for a month-long cruise in November. I also was asked to drive a zodiac boat to take the tourists from the ship to landings on the shore, because I had a small boat operator's license from working in Alaska (Fig. 4.19) I immediately accepted his offer, which turned out to be one of the most adventurous and best trips of my lifetime. This was an incredible

Fig. 4.19 **a** Photo of Hans Nelson in orange survivor suit next to Cheeseman Ecotours attendees and the black zodiac boat, which was used to transport passengers from the Russian oceanographic ship to the Antarctic Peninsula shore for a geological field trip led by Nelson in 1994. **b** Photo of Hans Nelson as lecturer on a Norwegian Cruise Line ship while visiting Glacier Bay in southeast Alaska in 2000. Photo sources are unknown

experience, because it was a chance to visit the most unchanged and natural area on our Earth.

Because this was an eco-tour, Doug Cheeseman had gathered many world-class scientists for the cruise, including ornithologists, marine biologists, Antarctic historians and some of the world's best nature photographers. These photographers knew that Cheeseman's eco-tour offered much better opportunities for photography than typical tours. They were correct because we spent much more time and had many more shore landings than any other previous Antarctic tour. Prior to our Cheeseman tour, the most zodiac landings any Antarctic tour made were 15. We almost doubled that and made 25 landings. Not only that, but instead of spending an hour or two on the shore, we sometimes spent as much as 20 h on a landing and hiked across a Falkland Island from one side to the other.

Our trip to Antarctica began in Ushuaia at the southern tip of Argentina. In 1994 this was a small typical polar town of 15,000 people, similar to what I had seen in Alaska and Siberia. I returned to Ushuaia in 2017 and was surprised to find that it was a large city of over 100,000 people and covered with new buildings under construction. The Antarctic tourism had boomed, but also the Argentine government had subsidized a number of new industries. Most important in 2017 I discovered that there was a large new scientific laboratory that had been developed for environmental studies in southern Argentina and Antarctica. These studies will help to document effects of any future global changes.

In Ushuaia we boarded a large Russian oceanographic ship. The Russians had been opportunistic to use their scientific ships during the Antarctic summer, when they could not be used as much during the Russian winter. Also as I mentioned previously, with the fall of the Soviet Union there was a huge collapse of Russian science and this was a way to keep their ships funded. The Russian oceanographic ships were two or three times larger than the typical United States oceanographic ships. The US ships normally could carry only 20 to 40 scientists and conduct scientific cruises for two weeks to a month, whereas the Russian ships could carry well over 100 scientists and conduct scientific cruises for several months to a year. Consequently, our Russian ship for the Cheeseman eco-tour could carry around 100 tourists, plus another 100 or so ship crew and scientists.

We traveled from Ushuaia to our first stop at the Falkland Islands as the British call their colony. Argentina names them the Malvinas Islands, and claim the islands belong to them, which resulted in a 1984 war with Great Britain. Our visit was a few years after the war and things were quite calm, although when invited in to visit Falkland Islanders homes, they described terrifying adventures when Argentina invaded the island. Fortunately, in 1994 things were back to normal and this was the first place we visited where we could see rock hopper penguin colonies. These were very amusing penguins as they lived up to their name by hopping up and down the cliffs along the shore. However, once they entered the sea they could swim incredibly fast and gracefully.

We went from the Falkland Islands to South Georgia Island, which was a major whaling center at the end of the nineteenth century. At this time the island also was famous because of the history of the Shackleton expedition to Antarctica [44]. After his ship was frozen in the Weddell Sea ice, Shackleton set off on a rescue mission to save his crew, which stayed with his boat near Egg Island along the Antarctic Peninsula. He crossed the extremely stormy seas of the Drake Passage to South Georgia Island in a small rowboat. When

Shackleton landed on South Georgia Island, he was on the opposite coast from the whaling village. After surviving the Drake Passage, Shackleton and his party had to climb over the glacial covered mountains to reach the whaling village.

When you experience the stormy seas of the Drake Passage between the Antarctic Peninsula and South Georgia Island, and see the high mountains of the Island covered with glaciers, the survival of Shackleton's rescue expedition was a miracle of leadership. The other miracle was that all the crew staying with the ship survived after waiting two years to be rescued. After reading books about the Shackleton expedition, it was quite moving when I went wandering around the cemetery at South Georgia and stumbled upon a small wooden cross where Shackleton was buried. Now apparently a large monument marks his grave. Another interesting aspect on our tour was that the British historian on our ship had some of Shackleton's original lantern slides, which he showed with his lectures.

When we visited South Georgia Island, it no longer was a whaling station, and most of the hunting of whales has ceased. One of the reasons that whaling has ended is that some of the whale populations have been nearly driven to extinction. The International Whaling Commission now has almost stopped the hunting of whales and is an example of an international effort to slow problems of human-caused global change. Unfortunately, the Japanese have put a stipulation into the Commission rules, that they could take whales for research. They abuse this privilege by killing 300 to 1200 whales in the Antarctic seas and Pacific Ocean, supposedly for research, but actually they are hunting them for food (https://www.bbc.com/news/world-asia-485 92682). There are ecological groups that physically try to stop the Japanese from hunting whales in Antarctica. In one confrontation with a Japanese whaling ship, the Japanese rammed the ship Sea Shepherd. When visiting Hobart, Tasmania in 2010, we could see the hole in the side of the Sea Shepard. The harassment has slowed down whale hunting by the Japanese for the supposed research.

There were some adventures related to South Georgia Island. The first was as we traveled along the eastern edge of the island on a beautiful sunny day, suddenly a white cloud came roaring down the mountainside. This was an example of katabatic winds, where at higher elevations, strong winds develop and accelerate down the mountainsides and often sweep up snow. In Antarctica the winds sometimes reach speeds up to 300 km (190 miles) per hour and are extremely dangerous (https://en.wikipedia.org/wiki/Katabatic_wind). The wind that came off South Georgia Island was not nearly as strong, but still it rapidly stirred up the sea and you could feel it hit the ship with force that lasted a few minutes.

Along the coast of South Georgia, we also took the zodiacs to visit a colony of penguins (Fig. 4.19). The captain was worried about the lack of good maps for the depth of the sea bottom, so he waited about 16 km (10 miles) offshore. Consequently, we had a long trip in the zodiacs, and as I started on the trip back to the ship, the motor on my zodiac failed. Normally this would not have been a cause for worry, but in this case we had a long transit and the sea was becoming rough. Without power you could not navigate the boat for the safest path through the waves. Luckily, another zodiac came along and hooked up a rope and pulled us safely back to the ship.

Another major adventure took place after we left St. George Island and traveled across Drake Passage to the Antarctic Peninsula. This sea has the world's worst weather and is often called the roaring 60 s. There is completely open ocean circulation with no landmass to slow the wind and as a result waves can build to the maximum height possible. In fact, if you look at a map showing the highest average wave heights at any location in the world, when you look at the roaring 60 s, the waves on average are the highest in the world. The roaring 60 s lived up to their name and we had the highest waves I have ever seen in my many years of scientific cruises.

We were on a large ship that was about as high as an eight-story building. However, when we hit waves, a huge sheet of water would come over the entire height of the ship. Normally the time to cross the Drake Passage should have been two days, but it took us four days because of being slowed down by the high waves. Because we had the travel time across the Drake Passage, I gave some of my geological lectures. I was very impressed by the interest of the tourists, because in spite of many of them being seasick, they filled the lecture room. When they became sick they ran outside, got seasick, and then returned back to the lecture. I realize that on the ship, I had a captive audience, but the eco-cruise was the best teaching experience of my life because of the 100% interest in science by all of the tourists.

The next adventure was when we reached the Antarctic Peninsula and visited Paradise Bay, one of the most beautiful places I've ever seen. After stopping to see several penguin colonies, we traveled through a narrow passage to reach the Bay that was enclosed by mountain glaciers traveling down to the sea. There were blue ice cliffs as much as 30 m (100) feet high reaching the deep blue-colored seawater. Huge blocks of ice were continually breaking off with a thunderous roar that echoed through the enclosed bay. The ice crashed into the sea and caused large waves to travel away from the impact and rocked our zodiac as we watched.

We also stopped and had a pleasant visit at the Argentine research station that was in Paradise Bay. These research stations from many countries were common on the Antarctic Peninsula. We also visited the Scott Polar Research station of Cambridge University and found it was staffed entirely by women scientists who each stayed alone in their snow-covered tents. These were adventurous young women that were undertaking important studies about penguin colonies. Another courageous pair, that we encountered one stormy day, was two Australian kayakers who were traveling and camping around the Antarctic Peninsula. This was extremely adventurous considering the katabatic windstorms that suddenly take place in Antarctica. They came aboard ship for a couple of days to get showers and share hot meals with us.

When we came into Paradise Bay, the narrow passage was completely open. When we returned from Paradise Bay, we found that the narrow inlet was completely blocked by ice. Fortunately, the Russia ship had an ice-strengthened hull and we slowly proceeded through the ice-choked passage. There was tremendous scraping and crashing of ice as we traveled the passage. When we looked back at the passage, we could see orange ice along the sides where the paint had been scraped off our ship. As we proceeded away from Paradise Bay, we visited other penguin colonies with millions of squawking, smelly penguins.

Our next big adventure was visiting Deception Island. This was a similar volcanic setting and a same size caldera as Crater Lake. In the case of Deception Island, however, this volcano collapsed into the sea and had a narrow passage that was eroded through a side of the caldera wall. This passage was just large enough for our ship to travel through and enter the circular bay that resided in the collapse caldera of Deception Island. It was fortunate that we could spend 20 h hiking around this volcanic mountain, because it was interesting for me to compare this caldera with the Mount Mazama caldera that contains Crater Lake, Oregon.

The Deception Island volcano has been active in the past few decades and still contained locations with steam vents and heated water along the shoreline. In this cold Antarctic location, we all swam in the hot water near the shoreline. However, one had to be careful where you stepped so that you did not burn your feet. Near the steam vents, we visited the abandoned British Antarctic Survey research station in Deception Bay. This was especially interesting because the British historian on our ship had copies of the original logbook that described the last major 1969 volcanic eruption on Deception Island. There was a minute-by-minute description of the explosions and volcanic debris that was raining down on the British station. A nearby Chilean research station was destroyed and these scientists escaped to join the British scientists. Fortunately, they all survived, but both the Chilean and British bases were destroyed by the eruption, and only remnants remain. This again highlights the many adventures that earth scientists go through during their research.

The other spectacular experience about the visit to Antarctica was the abundance of wildlife and the fact that the wildlife was generally not bothered by our presence. You could walk among the millions of penguins and they did not flee. They would just keep holding their eggs or young between their feet. If they had not yet laid eggs, you could watch males slowly picking up stones to build a nest and hopefully attract a female penguin. It was amusing to watch because sometimes one penguin would gather some rocks and be putting them in the nest, but while he went to gather another rock, a nearby penguin would snatch the first penguin's rocks to build his nest. It was amazing to observe how a penguin after going to sea to obtain food for their young chick, could then find their own chick among the other thousands of penguins in the colony. It also was interesting to walk next to albatross nests and realize that they may travel for weeks to obtain food and then return to feed their young. Some of these albatrosses had tracking devices and ornithologists learned that they might travel from Antarctica to the Arctic Ocean on a single feeding journey.

You certainly could view nature in the raw and typical wild biological interactions in a world with little global change. For example, skuas, a type of hawk, were constantly swooping about the penguin colonies and sometimes came down next to you to snatch a penguin chick or an egg. I also observed penguins that jumped into the sea and immediately were attacked by leopard seals, which knocked the penguin into the air and then caught and killed it. I also saw penguins that had huge chunks bitten out of them by fur seals.

Male fur seals were extremely aggressive during the breeding season and would attack anything that came near them, such as a penguins or humans. We had to be extremely careful to stay far away from fur seals. The year before a tourist ventured too close to a seal and suffered a bite that took 50 stitches to close. Apparently this breading time is extremely stressful because only a small percentage of the male fur seals collect a harem of females. Most male seals do not breed and only live half as long as the female seals. One also did not want to get too close to the elephant seals that were in the breeding season. They were comical to watch because the young male seals would lumbar at each other and bang their heads together until they became all bloodied and were too tired to fight anymore. Meanwhile other successful older and larger breeding elephant seals lay about enjoying their harems of many females.

I already mentioned one zodiac adventure with killer whales in the Antarctic. I had other adventures because as a zodiac driver I took the zodiac to do my own explorations (Fig. 4.19). I became particularly interested in the beaches, which were unlike beaches outside Polar Regions. In the polar beaches a wide variety of rocks are present, unlike beaches in other regions where the rocks are the same as the nearby rocky headlands. The presence of all the exotic rocks in polar beaches results because as the ice freezes to the bottom, it picks up pebbles in the ice. The ice can move away from where it picked up the rocks and travel far away to another beach and drop rocks that were carried from another location. Consequently, a wide variety of exotic rocks are found in the Antarctic beaches. As part of my investigation I would drive the zodiac through partially melted icebergs and observe the rocks that had been picked up by the ice. I used these methods until one day when I saw an iceberg overturn with a huge crash because most of the iceberg of course is below the water. After that I no longer drove my zodiac through the icebergs because of the danger of them turning over.

4.13   Earthquakes and Hazards

As I pointed out in the preface, we humans live in a natural world and are subject to natural global change as well as our human-caused changes. My geologist journey has focused on both of these types of changes and their affects on humans. The natural changes impact people most immediately, such as catastrophic loss of life from natural disasters, which has averaged 53,000 per year for the last 30 years (USA Today, January 8, 2019). In contrast, the human caused changes are slower to affect humans. However, because of the increasing dominance of humans over our entire global environment, the human caused changes will have the greatest effect on the human race in the long run. Already, millions of lives are lost to air pollution (see Chapter 5 Sect. 6.4), and economic costs are increasing dramatically. Previously, the average global cost of natural disasters has been about 140 million dollars per year (USA Today, January 8, 2019). However, the average cost for the past two years has nearly doubled because of worse extreme weather events that appear to be related to global warming (see Chapter 5 Sect. 5.7).

I have concentrated my research on the natural global changes from volcanic eruptions and earthquakes because they cause the most immediate catastrophic loss of life. Previously I described my global change research related to Crater Lake. In terms of loss of life, volcanic eruptions are not nearly as catastrophic as earthquakes and related tsunamis. However, great volcanic eruptions can cause more significant natural global change that can affect climate for centuries. The largest historic loss of life from volcanic eruptions ranges from 3,500 to a maximum of 71,000–250,000 lives from the Mount Tambora eruption in Indonesia in 1815 (https://en.wikipedia.org/wiki/List_of_volcanic_eruptions_by_death_toll).

Five of the 10 most historic catastrophic losses of life from natural geologic phenomenon are related to earthquakes and the number of deaths ranged from 280,000 to 830,000 (https://en.wikipedia.org/wiki/List_of_natural_disasters_by_death_toll). Since 1900, five of the ten most catastrophic losses of life (143,000 to 280,000 lives) again resulted from earthquakes. Two of these five earthquakes have occurred since 2003. The recent large loss of life from earthquakes, in part, is related to population increases and greater numbers of people living near the coast in tsunami zones. Since 1995, I have been undertaking earthquake research on Cascadia and northern San Andreas fault earthquakes and also with European scientists on the Crete 365, western Sicily 1169, Catania 1693, Lisbon 1755, and Messina 1908 A.D. earthquakes, which had losses of life ranging from 50,000 to 123,000 people in each earthquake (Fig. 2.8) [43].

In 1995 I began a new project to define the earthquake history of the Cascadia Subduction Zone because of my continuing studies on the Cascadia continental margin offshore from the Pacific Northwest (Fig. 2.8). This history is based on the deep-sea seismo-turbidite sand layers deposited by turbidity currents that are triggered by earthquakes (see the dark gray sand layers of the 1999 Cascadia box core in Fig. 3.3b). When the world's largest Mw 9 earthquakes occur, they shake continental margin slopes for distances of more than 1,000 km (see slope in Figs. 2.8, 4.20), whereas smaller earthquakes shake slopes for shorter distances. As a result, synchronous multiple submarine landslides evolve into multiple turbidity currents (Fig. 4.20). These currents travel down canyons and into channels on the deep seafloor and deposit episodic sand layers that are evidence of the earthquake (Figs. 2.7, 2.8). We can find the date of the earthquakes with radiocarbon ages from the slowly deposited mud between the sand layers (Figs. 2.7b, 4.21) [45]. However, this is not a simple process where you just take some cores of sediment in the deep seafloor and obtain ages for an earthquake.

To obtain earthquake history, I developed new detailed research methods and then cooperated with Dr. Chris Goldfinger at Oregon State University in the late 1990s. I will describe these techniques in the following paragraphs, but the most important value of these new methods is that we can obtain the frequency and recurrence time between earthquakes for thousands of years,

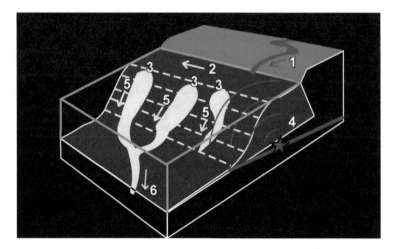

SEQUENCE OF EVENTS

1. Rivers deliver sand to the shelf
2. Sediment is transported along shelf
3. Sediment accumulates in canyons heads
4. Earthquake occurs & shakes margin
5. Sediment funnels down canyons-channels as turbidity currents
6. Turbidites deposit in canyons and channels to create a paleoseismic record
7. Earthquake turbidites correlate in time with land tsunami sands

Fig. 4.20 Top diagram shows how seismo-turbidites are formed by great earthquakes (Mw 7 to 9) and deposited on ocean or lake floor basins. Synchronous triggering of turbidity currents in multiple submarine canyons provides evidence that a seismo-turbidite has been triggered by an earthquake (see discussion in Sect. 4.13). With a history of multiple seismo-turbidites and the time between them, the earthquake history and hazard of an earthquake fault can be determined (e.g. Figure 4.21). Drawing is modified from Adams [48] by Brian Atwater, U.S. Geological Survey, Vancouver, WA

compared to a shorter history from coastal studies of tsunami sand layers and drowned forests (Fig. 4.21). With this longer history of deepwater sand layers, you can assess the earthquake hazards better and determine the timing plus history of unusual super quakes like 2011 in Japan [46]. Using our methods, hundreds of scientists have initiated studies of earthquake history and hazards around the globe so that this history can be used to develop safe building codes [47].

With this introduction, I will provide some background about how our new studies of deep-sea earthquake records evolved. Little did we realize during the 1960s, that our numerous PhD studies in the oceanography

CASCADIA SUBDUCTION ZONE

NORTHERN SAN ANDREAS FAULT

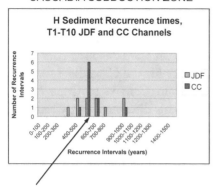

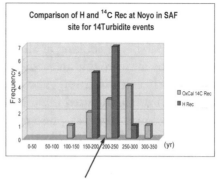

Fig. 4.21 Histograms of Earthquake recurrence times for the Cascadia Subduction Zone and the northern San Andreas faults based on the time between seismo-turbidites. Earthquake recurrence times for Cascadia are based on hemipelagic clay thickness between 10 seismo-turbidites in Juan de Fuca and Cascadia Channels. Earthquake recurrence times for the northern San Andreas Fault are based on both hemipelagic clay thickness (13 events) and radiocarbon ages (11 events) in Noyo Channel off northern California. The Cascadia histogram shows that the average time between Mw 9 great earthquakes on the Cascadia Subduction Zone for the past 10,000 years is 530 years (see arrow). The minimum time is about 300 years and the maximum time is about 1000 years [51]. The San Andreas histogram shows that the average time between Mw 8 great earthquakes on the northern San Andreas Fault for the past 10,000 years is about 200 years (see arrow). The minimum time is about 100 years and the maximum time is about 300 years. Note that both fault systems have entered the minimum estimated earthquake recurrence time between great earthquakes because the last great earthquake in Cascadia was 1700 A.D. [72] and the last great earthquake on the northern San Andreas Fault was the 1906 San Francisco earthquake. Figure source is Gutierrez et al. [45]

department at Oregon State University would become valuable for earthquake history. At that time, our thesis advisor Dr. John Byrne developed a logical research strategy of having his students work progressively from the estuaries, to the shoreline, then to the continental slope and then to the deep-sea Cascadia Basin. Because of this complete coverage of studies on the entire underwater Cascadia margin and the uniform methods, our PhD results could be combined later on by Dr. John Adams of the Canadian Geological Survey [48]. Dr. Adams had a unique interdisciplinary background, because he did his PhD study on deep-sea sediments in New Zealand, with our Antarctic scientist friend Dr. Peter Barrett. When Dr. Adams began working with the Canadian Survey, he changed his research to paleoseismology, which is the study of the past history of earthquakes.

Initially in the early 1980s, Dr. Brian Atwater of the USGS undertook shoreline studies of drowned forests and tsunami deposits that indicated subduction zone earthquakes had taken place along the Cascadia continental margin off the Pacific Northwest of North America [49]. Then, because of his combined skills of sediment studies and paleoseismology, Dr. Adams in the late 1980s began looking at all of our Oregon State deep sea core records to see if they may provide an earthquake history for the Cascadia Subduction Zone, which extends from Vancouver Island in Canada to Cape Mendocino in California. He used my initial PhD thesis discovery that Crater Lake Mazama volcanic ash was present in the sand layers of Cascadia Basin in the deep sea off the Pacific Northwest (Fig. 2.8) [4]. The first occurrence of Mazama ash in a sand layer provided a geologic time marker with a known age for the 13th sand layer in the sediment of each core.

Dr. Adams [48] noted that cores in most of the channel systems along the 1,000 km Cascadia margin had 12 sand layers above the initial occurrence of Mazama ash in the 13th sand layer down from the seafloor surface. From this he made the brilliant observation, that if every channel system fed by submarine canyons had the same number of 13 turbidite sand layers above this known age of the Mount Mazama volcano eruption about 7600 years ago, it must be a record of earthquakes that took place on average about every 500 to 600 years. This same sequence of turbidite sand layers everywhere along the Cascadia margin could only be caused by the entire 1000 km (620 miles) of the margin being shaken synchronously by a Mw 9 earthquake to trigger the same number of sand layers at the same time at every channel location along the margin (Fig. 4.20).

After Dr. Adams published his article in 1990, I began reviewing my PhD thesis studies. In contrast to Cascadia Channel, where all the away down the channel there were 13 post-Mazama sand layers [48], 150 km down Astoria Channel I found only three post-Mazama layers (Fig. 2.8) [4]. Consequently, I questioned whether Adams theory was correct. At about the same time in the mid-1990s, I was giving a poster of my results at the American Geophysical Union meeting in San Francisco. Fortunately, next to me was Dr. Chris Goldfinger, who also questioned Adams theory about the 13 sand layers. We began talking and realized that we had the ideal combination of skills to work on this problem of the Cascadia earthquake history. I was specialized in the detailed characteristics of bottom sediments and Goldfinger was specialized in the study of faults, generation of earthquakes, and seafloor mapping.

I continued working on my earthquake project in Cascadia Basin at the USGS and proposed studies to collect more cores of sediment to prove or disprove Adams theory. However with the continuing cuts in funding for the USGS and no ship, it was impossible to have a new cruise to collect the necessary cores of sediment. Finally, in 1998 I retired from the USGS, because it was impossible to undertake these critical new studies for the USA national need to define earthquake hazards. Once I retired, I could work with Dr. Chris Goldfinger at Oregon State and try to obtain funding for our proposed study through the USA National Science Foundation. I gave Dr. Chris Goldfinger my USGS proposal and he submitted it to the Marine Science Division of the National Science Foundation (NSF). They refused to fund our proposal, so we submitted it to the Tectonics Division of the NSF. They immediately saw the value of this interdisciplinary study. In fact, two of the reviewers for NSF said it was one of the best proposals they had ever reviewed. Consequently, we were funded for new scientific cruises in 1999 and 2002 to obtain more cores to define the Cascadia Subduction Zone earthquake history.

For our 1999 cruise, I recruited a number of students and had several from different European countries. I was particularly impressed with the excellent training of the European undergraduate students, compared to the American students. For example, the Spanish students had five years of geology courses to complete their undergraduate university degrees. American students, including myself, typically had only about two years of geology courses and only one course in each type of geology. In contrast, the Spanish students and other European students had two or three courses in each specialty of geology.

As a result, the foreign students were able to accomplish some of the sophisticated research that we undertook on board the ship. They could identify the microscopic shells that we needed to determine ages of the sediment. We required this research on board the ship, because we opened the cores and did some analyses of the sediment on the ship rather than in laboratories onshore. One of the important tasks on board the ship was to identify where we first found the occurrence of the Mazama ash, so that we could determine if we had 12 sand layers above the first occurrence of sand with Mazama ash (e.g. Fig. 2.7b). We needed to do this on the ship so that we knew we had obtained the best core record possible at each channel location to compare with Dr. Adams results. A second important task was to determine the other time marker, which was the change from the ice age Pleistocene time sediment to the warmer climates of the Holocene time.

Before doing the critical task of finding the time marker of the Mazama ash in the sediment cores, we had to do a detailed analysis of the seafloor swath bathymetry for the best location of coring sites in submarine channels Fig. 2.5, 2.8. To determine the location for a core site, we innovated new techniques that had never been done at sea. We had an integrated GIS base of all the swath bathymetric information in Cascadia Basin. Consequently at sea, we could obtain new swath bathymetry and integrate it with previous information to obtain the most accurate map of the seafloor channels With these techniques we could then use computer software called Fledermaus and fly through the GIS database with the new data and see the detailed shape or morphology of the seafloor, such as the channel pathway or small terraces beside this pathway (e.g. Fig. 2.8). The morphology was critical to obtain the best possible location to obtain a complete and accurate record of the 13 post-Mazama sand layers.

Our methods were to carefully locate a number of cores in transects across and along channels (e.g. Fig. 2.8), and then open the cores and be sure at each location that we found the most complete record of the 13 post Mazama sand layers (Fig. 2.9b). In this way we could prove or disprove John Adams theory. The cruise was successful, because in the best core at every site, we found the 13 sand layers [5, 6]. We also solved my problem of observing only three post-Mazama sand layers in distal Astoria Channel on Astoria submarine fan. We observed, that the further down we went on Astoria Channel, the fewer post-Mazama sand layers we found. Consequently, the distal Astoria Channel did not have as complete a record of the 13 sand layers similar to the other types of channels like Cascadia Channel. In summary, we proved Dr. Adams theory that counting up the 13 sand layers provided a good earthquake history for the Cascadia Subduction Zone and that the 13 sand layers were seismo-turbidites generated by earthquakes [48, 5, 6].

Once we had confirmed that we had a valid earthquake history, we began detailed studies to determine how frequent these earthquake-generated sand layers (seismo-turbidites) were deposited. This was the most important part of our study, because it was the way to determine the earthquake hazards for the entire Pacific Northwest coastal area. These studies consisted of examining the continuously deposited fine mud (hemipelagic mud) between the episodically deposited sand layers in all our cores (Figs. 2.7b, 3.3b, 4.21) [45]. The hemipelagic mud thickness was measured and the microscopic foraminifera shells originating from the sea surface were collected for radiocarbon ages. These sea surface shells sink to the deep seafloor mud and provide the most accurate radiocarbon ages of the hemipelagic mud.

To determine earthquake frequency from how often the seismo-turbidites were deposited, we used two independent methods. One method, the H method, was to determine how thick the hemipelagic mud layer (H) was between each episodic seismo-turbidite (Figs. 2.7b, 3.3b) [45]. The second method was to use high-resolution accelerator mass spectrometer radiocarbon ages of the foraminiferal shells in the hemipelagic mud beneath each seismo-turbidite [47]. By using these two methods to determine the time between each seismo-turbidite, we were then able to find what the minimum time, average time and maximum time was between (Mw 9) great earthquakes (Fig. 4.21) [45]. The most important thing to determine was the minimum time between great earthquakes during the past 10,000 years. This would give the best information for earthquake hazard planning for the Pacific Northwest. For example, if the minimum time between two great earthquakes was several hundred years and the last great earthquake was 10 years ago, there would not be a significant earthquake hazard for several hundred years.

To determine the time between earthquakes with the H method, the thickness of the hemipelagic mud (H) is measured below each seismo-turbidite (Figs. 2.7b, 3.3b, 4.21) [45]. The H mud is the normal continuously deposited material that rains down from the water column. It contains land-derived suspended sediment and dust, plus the microscopic shells of plants and animals living in the water column. It deposits at a very constant rate of about 10 cm (4 inches) per thousand years over the deep seafloor in Cascadia Basin [50]. Consequently, in Cascadia basin when there is 3 cm (1.2 inches) of H mud between two seismo-turbidites, the time between earthquakes is 300 years. We find that during the past 10,000 years about 3 cm (1.2 inches) is the thinnest H mud between seismo-turbidites, the average is about 5 cm (2 inches) and the maximum is about 12 cm (5 inches). From this we conclude that the minimum time between Cascadia great earthquakes is about 300 years, the average time is about 530 years and the maximum time between two earthquakes is about 1000 years (Fig. 4.21 Cascadia) [45]. As mentioned previously, the last Cascadia great earthquake took place in January 1700 and it is now over 300 years since this last earthquake. Thus the Pacific Northwest is in the very early window for another possible Cascadia Subduction Zone earthquake. Using earthquake hazard risk analysis, this indicates that there is a 7–12% chance of a Cascadia great earthquake within the next 50 years [47].

The second method to determine time between great earthquakes is to obtain radiocarbon ages from above and below each seismo-turbidite sand layer (Fig. 2.9b, 3.3b). To do this, the H mud immediately above and below each sand layer is sampled and then the microscopic sea surface planktonic foraminifera shells are separated out. These small amounts of carbonate shells can have their high-resolution radiocarbon ages determined from the carbon in the shells. Once we determine the radiocarbon age for each hemipelagic mud layer, the time between two consecutive seismo-turbidite can be obtained by subtracting the radiocarbon age of the younger H layer above from the older H layer below the seismo-turbidite. When this is done, the radiocarbon age method independently confirms the same result for minimum, average and maximum time between earthquakes that we determined by using the H method (Fig. 4.21 San Andreas) [45, 47].

The relative strength of earthquakes is another important aspect to determine. By comparing the heights of Cascadia tsunami wave deposits with the tsunami wave heights and strengths of historic earthquakes, we think that the 13 post-Mazama Cascadia earthquakes result from great earthquakes of around Mw 9 strength [51]. These estimates of earthquake strength and frequency apply to the northern Cascadia margin that extends from Vancouver Island to central Oregon. We find that in southern Oregon and northern California, there are weaker earthquakes in addition to the 13 Mw 9 quakes, and earthquakes become more frequent towards the Mendocino triple junction where the Cascadia Subduction Zone and the northern San Andreas faults meet [47]. Consequently, there is a 21% chance of an earthquake in the next 50 years in the southern Cascadia area, whereas there is only a 7–12% chance of a Mw 9 earthquake in northern Cascadia during the next 50 years (Fig. 4.21).

After having success with our studies of the Cascadia Subduction Zone earthquake history, we decided to apply our methods to study the earthquake history of the northern San Andreas Fault. This fault extends from Monterey Bay in central California to the Mendocino triple junction in northern California. This famous fault traverses along the coastline through San Francisco and eventually offshore where it cuts the Noyo submarine canyon head. We again surveyed a number of submarine canyons and channels such as Viscaino, Noyo, and Gualala from the north towards the south. After surveying the canyons we took deep-sea cores to examine the history of seismo-turbidites. Similar to Cascadia basin, we determined the radiocarbon age of each seismo-turbidite sand layer and the time between each sand layer generated by a great earthquake along the northern San Andreas Fault (Fig. 4.21 San Andreas).

We find that the average time between 30 of these seismo-turbidites for the past 10,000 years is approximately 200 years (Fig. 4.21 San Andreas) [45]. The minimum time between two seismo-turbidites is approximately 100 years and the maximum time is approximately 300 years. Consequently, like Cascadia Basin, the northern San Andreas Fault is in the early time window for another earthquake. We estimate it is about 100 years on average before another earthquake may take place on the northern San Andreas Fault, because the 1906 San Francisco earthquake took place over 100 years ago and the average time between earthquakes is 200 years. These earthquake frequencies for the northern San Andreas Fault, however do not apply to be southern

San Andreas Fault. The southern fault is much more complex because it has many different fault segments that can create earthquakes.

Several important observations can be made by comparing the earthquake frequency in the Cascadia Subduction Zone with that of the northern San Andreas Fault zone (Figs. 2.9b, 4.21). The first observation is that these different earthquake frequencies provide a good history for these two different fault zones. Some experts have questioned whether deep sea sand layers can provide an earthquake history, because there are numerous processes besides earthquakes that can generate the deposition of deep-sea turbidite sand layers of the type that we studied for earthquake history. These other processes include major cyclonic storms, random submarine landslides, river floods introducing large plumes of sediment into the sea and tsunamis from other earthquakes across the ocean that wash shallow water sediment back into the sea in the Cascadia area.

In the case of Cascadia Subduction Zone and the northern San Andreas faults, all of these other possible processes for generating turbidite sand layers can be eliminated because the two faults meet at the Mendocino triple junction. Any of these other storm, flood or tsunami processes, that could trigger turbidites, would affect both areas where these two fault systems meet. Consequently, the other possible triggering processes could not be generating the turbidite sand layers we study or there would not be two different frequencies of sand layer deposition on either side of the Mendocino triple junction where these two fault systems meet [51].

The average earthquake frequency on the north side of the triple junction is about 530 years, whereas on the south side the frequency is about 200 years (Fig. 4.21) [45]. The only way you can create these two different frequencies of turbidite sands in the same area is by different earthquake generation times on the two different fault systems on either side of the triple junction. In addition, the onland paleoseismic records for Cascadia and San Andreas earthquake are different and agree with our seismo-turbidite ages for earthquakes [52, 47]. This proves that seismo-turbidite sand layers provide a valid earthquake history record for both the Cascadia and San Andreas faults.

The Cascadia and San Andreas Fault systems have two different frequencies of earthquakes because they have two different types of faults and the length of earthquake rupture on the faults is different. When the entire 1000 km (600 mile) length of the Cascadia Subduction Zone ruptures it creates Mw 9 earthquakes and these only occur about every 530 years (Figs. 2.9b, 4.21 Cascadia) [45, 47]. In contrast, maximum rupture length of the different northern San Andreas transform Fault is only about 300 km (190 miles) and causes earthquakes of only about Mw 8, and these earthquakes occur about every 200 years on average (Fig. 4.21 San Andreas). This shorter earthquake rupture length results in an earthquake generating strength that is nearly 1000 times less for northern San Andreas earthquakes.

It appears that we can see this difference of earthquake strength in the deposits of seismo-turbidites that are generated by these two different faults. With the world's strongest Mw 9 earthquakes that are created on subduction zones like Cascadia, we see that the great strength of these earthquakes shakes the continental margin with such force that the seismo-turbidites exhibit sand pulses that are related to the way the subduction zone fault rips apart or ruptures. These sand pulses are coarser-grained sand units within the single seismo-turbidite created by an earthquake event.

The 2004 subduction zone earthquake off Sumatra had three major ruptures and the seismo-turbidite generated by this earthquake has three sand pulses [53]. Similarly, the 2011 Japanese subduction zone earthquake had two ruptures and the seismo-turbidite has two pulses [46]. In Cascadia we see that each seismo-turbidite has a characteristic set of sand pulses for each earthquake event in every channel along 1000 km (600 miles) of the Cascadia margin [47]. Likewise, every seismo-turbidite from each earthquake has a similar thickness in all channels along the margin. Because this thickness varies for each earthquake, it appears that the seismo-turbiditethickness indicates different strengths of earthquakes.

For example, the 11th seismo-turbidite below the seafloor surface in Cascadia Basin is always the thickest and apparently represents a larger earthquake than normal [47]. By looking at these abnormally thick seismo-turbidites and larger tsunami sand deposits, Goldfinger et al. [46] have been able to develop a theory of less frequent super quakes or anonymously large earthquakes for a fault system. They see super quakes in the Cascadia Subduction Zone (e.g. layer 11) and also in the Japanese subduction zone of the 2011 earthquake. We now know that Mw 9 earthquakes take place about every 800 years in the area of the Japanese 2011 earthquake, whereas previous studies indicated that only magnitude 8 earthquakes occurred in this area. These results show the value of the longer earthquake history that we can obtain from the deep-sea seismo-turbidites compared to the shorter history of tsunami sand deposits in coastal records (e.g. [49, 47].

Another important result of comparing the Cascadia and northern San Andreas earthquake history is, that after a Cascadia Mw 9 earthquake, about 70% of the time the northern San Andreas Fault will have an earthquake within 20 years [52]. This is the first time in the world that one earthquake system has been linked with another earthquake system. This is not surprising, because when a Mw 9 earthquake occurs, it stresses the Earth's crust for about 100 km distance from the fault epicenter. Consequently, in the setting where the Cascadia Subduction Zone fault extends to the northern end of the San Andreas Fault, the northern end of the San Andreas Fault is stressed for about 100 km (62 miles), or one third of its length. As a result, about 70% of the time, a Mw 8 earthquake results on the northern San Andreas Fault after a Cascadia Mw 9 earthquake.

We have continued to expand our studies to other areas worldwide, such as the Ionian Sea and the Sumatra earthquake of 2004 [43, 53]. We have obtained significant results, such as in the Messina Strait area in southern Italy, where there have been three historic earthquakes in the past several hundred years (AD 1908, 1693, 1160) (Fig. 4.22) [43]. Each of these earthquakes has killed tens of thousands of people along the coast, such as the 1908 earthquake, which killed over 60,000 people.

My work with the Italian scientists has not only highlighted these earthquake hazards, but we also have been able to identify all the processes and

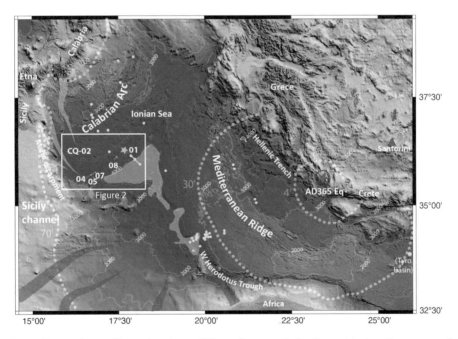

Fig. 4.22 A three dimensional rendition of a swath bathymetric location map of the Eastern Mediterranean Sea. This map shows the Ionian Sea basin, Crete AD 365 earthquake site, HAT seismo-turbidite deposits (irregular gray areas on sea floor), multiple submarine canyon pathways (in brown arrows with points into the seafloor) that transport earthquake generated turbidity currents, and location of Cala 04 core inside Fig. 2 box. Messina Strait is located between Etna and Calabria. Figure source is Polonia et al. [43]

seismo-turbidite deposits related to the last three historic earthquakes in the confined basin of the Ionian Sea (Fig. 4.22). At the base of each earthquake's deposit, we observe multiple sand layers stacked one on top of the other (Fig. 4.23 ST1a). These layers (St1a1, St1a2, St1a3 in Fig. 4.23) result from multiple turbidity currents that an earthquake synchronously triggered in different canyons along the Italian margin (Fig. 4.22). Because there

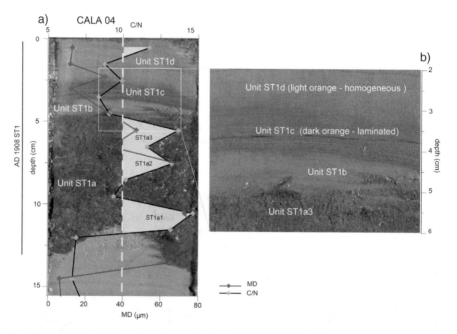

Fig. 4.23 Part **a** Photo showing seismo-turbidite ST1 in piston core CALA 04. ST1 was deposited on the Ionian deep-sea floor off southern Italy by the 7.24 Mw 1908 Messinian earthquake. The MD (red line) shows the size of sediment particles and C/N organic carbon to Nitrogen ratio (black line) shows more nearshore input of sediment particles. C/N ratios greater than 10 are marked by yellow areas. The yellow areas and the red line of larger sand grains identify the stacked ST1a1 to ST1a3 seismo-turbidite sand layers created by synchronous submarine landslides and turbidity currents triggered in different canyons (see Fig. 4.22 for canyon locations) by the Messinian 1725 AD earthquake The yellow box indicates the location of the enlarged photo in part b. Part **b** Enlarged photo of the upper part of ST1 in core CALA 04 shows the stacked sand unit ST1a 3 from a submarine landslide and turbidity current, ST1b homogenous finer sediment deposited from suspension, ST1c white and dark orange millimeter-thick laminated sediment from seiches (water sloshing back and forth) of the Ionian Sea for two weeks after the earthquake, and ST1d of nearshore sediment backwashed from the coast by the earthquake tsunami. The deposits in multiple cores from this 1908 earthquake and other earthquakes reveals that the entire processes and history of an earthquake event can be documented in the seismo-turbidite deposit of a confined basin like the Ionian Sea (Fig. 4.22). See Fig. 4.22 for core CALA 04 location. Figure source is from Polonia et al. [43]

are multiple turbidity currents, large clouds of suspended sediment in the confined basin slowly deposit a homogenous silt layer on top of the basal sand layers (Fig. 4.23 ST1b). Above this we see cyclic thin silt lamina that we think are related to the sea water sloshing back and forth or seiches of water that are caused by the earthquake shaking and tsunami in the confined Ionian Sea basin (Fig. 4.23 ST1c). Seiches have been observed in confined lake basins for up to two weeks after earthquakes in Switzerland and also after the 1908 Messinian earthquake [54, 43]. The fourth layer in this sequence of deposits, caused by one earthquake event, is a final cap of silt and clay that contains shells and carbon from shallow water locations. We think this final layer results from the material that is washed back into the sea from the coast by tsunami waves from the earthquake (Fig. 4.23 ST1d) [43].

Figures 4.23 and 4.24 provide a summary of the seismo-turbidite deposits and depositional processes resulting from a single earthquake in a confined basin. If this sequence of deposits can be identified in a rock outcrop, the ancient rocks can be interpreted to result from an earthquake (e.g. [55]. Identifying rocks from an earthquake can help a geologist understand the past geologic history of an area.

Using the Ionian seismo-turbidites to analyze the present earthquake hazards of the eastern Mediterranean Sea is even more important than defining past geologic history. A huge thick earthquake deposit that is found throughout the eastern Mediterranean Sea has intrigued many scientists. My colleague, Dr. Alina Polonia, has shown that this deposit is related to an estimated Mw 8.3 earthquake off the island of Crete in 365 A.D (Fig. 4.22) [42]. Historical records show that this earthquake destroyed many of the cities around the eastern Mediterranean, such as Alexandria, Egypt. An example of this destruction in Alexandria is descriptions of boats washed up on top of houses. We now are using our evidence from the more recent Messina earthquakes to better understand the hazards from a Crete type great earthquake where tsunamis in the future can potentially impact the coastal area of the entire eastern Mediterranean Sea.

Our work on the Cascadia Subduction Zone and San Andreas Fault has not only been utilized by scientists in other areas, but it also has attracted a great deal of public interest and been reported in several books on earthquakes. The Cascadia study was a featured article in the New Yorker magazine in July 2015, and this article won a Pulitzer Prize. Our research also has won the Geological Society of America Kirk Bryan award for research excellence in 2016. This award is sometimes called a Nobel Prize for studies of earth history during the past 2 million years. This certainly has been the crowning achievement of my nearly 60-year career in geology.

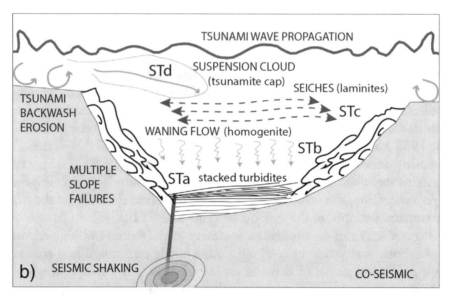

Fig. 4.24 Schematic diagram shows the sedimentary processes that result from (1) seismic shaking of an earthquake (see b) and (2) propagation of a tsunami wave generated by seafloor offset along a fault and/or resulting from slope failures caused by earthquake shaking. The tsunami wave may interact with the seafloor fault offset and slope failures to generate water sloshing of the entire basin water (i.e. seiches). The basin floor deposits (STa to STd) that result from the earthquake shaking, tsunami and seiches in a confined basin can be observed in Fig. 4.23. The first deposit of multiple sand layers (STa stacked turbidites) are related to multiple earthquake-triggered slope failures that cause multiple turbidity currents and deposition of stacked sand/silt layers at the base of the seismo-turbidite. The second deposits are homogenite soupy mud (STb) deposited on the deep sea floor from the suspended sediment of the multiple waning turbidity currents. Seiching of confined basin water masses reworks seismo-turbidite homogenites and deposits dark orange-colored laminated clayey silts (STc). Seafloor erosion of shoreline and shelf by tsunami waves washes back offshore, flows into the basin and creates slow sedimentation of final suspension cloud that deposits graded clayey silts and silty clays (STd). This general process model and the typical core deposits (see Fig. 4.23) for a seismo-turbidite (ST) sequence applies to confined basins where seiching can occur (i.e. lakes, fjords, Ionian and Marmara seas). Figure source is Polonia et al. [43]

The results from our earthquake hazards research are also significant for global change problems. Our findings, do not allow us to predict a specific time and place where an earthquake may take place, and most likely this will never be possible. However, our long history of earthquakes in a region can be used to protect global populations from earthquake catastrophes. We now know how often and the minimum time between great earthquakes in the Cascadia Subduction Zone and along the northern San Andreas faults. Good building codes have been enacted, particularly for San Francisco, because of the San Francisco earthquake in 1906.

Because we have only verified the Cascadia earthquake history during the past two decades, little has been done to prepare infrastructure for the next Mw 9 earthquake. Such an earthquake has the potential to cause thousands of deaths and trillions of dollars of property damage in the Pacific Northwest. This lack of preparedness for a Cascadia earthquake is pointed out in the 2015 New Yorker article. An important point made by the New Yorker article is that scientists are providing the factual knowledge about hazards, but the public interest, political will and funding has not developed to prepare for them. Recent documentaries show how little has been done to upgrade the infrastructure in the Pacific Northwest (e.g.https://www.youtube.com/watch?v=g3NlCXA6_kE). For example, it is estimated that 1700 bridges in the Pacific Northwest need to be retrofitted for a great earthquake, but only one of them has been retrofitted so far. Even this retrofit only assures that the bridge will not collapse and kill people, but it will probably be destroyed in a Mw 9 earthquake. In contrast, all Japanese bridges are retrofitted to continue to function and not be destroyed by the strongest earthquake shaking.

Unfortunately, the efforts to upgrade building codes and infrastructure along the San Andreas Fault of California have not been fully developed in the Pacific Northwest, although building codes are being revised and the legislature in Oregon is now providing some funding for infrastructure. Still documentaries show examples where some Oregon schools have not been moved from the tsunami zone and some new hospitals are being built in the tsunami zone. We scientists are frustrated that our studies on natural global change hazards are not being fully utilized to prevent loss of life and damage to infrastructure.

There are other simple things that can be done globally to prevent catastrophic loss of life. Although the Pacific Northwest has had little preparation of infrastructure for a great earthquake, they have an excellent warning system along the coast for tsunamis. There are signs all along the coast about how to escape a tsunami and there are warning sirens when a tsunami may occur. Similar systems could be installed around the world to prevent the catastrophic loss of life that results from tsunamis in coastal areas. At a minimum, basic education about tsunamis could be spread throughout the global coastal areas where great earthquakes occur. This would cost little or nothing and could be compared to the common education of California school children, where each school teaches children the basic protection of duck and cover during an earthquake. For tsunamis, children and adults worldwide should be taught that when they see the sea suddenly withdrawing far offshore, they should run to higher ground.

4.14   New Millennium Studies

During my retirement in the new millennium, my research continued on all phases of earth science related to global change, including resources, hazards and environmental assessment. Previously, I mentioned the continuation of my studies related to global earthquake hazards. In 2002, I began working in Spain on Antarctic climate history and its relation to ocean currents around Antarctica. The Antarctic currents control the entire world ocean circulation and thus drive the world climate system. My studies began with Spanish scientists working on the ocean circulation in the Bansfield Strait and Weddell Sea areas [56]. Detailed studies of the seafloor surface and subsurface were undertaken to determine the circulation patterns.

These studies used swath bathymetry, sidescan sonar and high resolution seismic profiling to outline the seafloor morphology and topographic characteristics (Fig. 2.5). Directions of currents at present and in the past could be determined by looking at the different patterns of sediment waveforms on the seafloor and in the subsurface [57]. An important aspect to determine was when did Antarctica separate from South America so that the Antarctic circumpolar currents could spin around the globe uninterrupted by landmasses. These currents are important for the world climate because deep cold water generates around Antarctica and then the circumpolar current circulates and introduces the cold Antarctic deep water to the world's oceans. This sets up a global circulation of ocean currents and climatic influence.

The other important part of global climate history is when did the Antarctic climate cool to form massive continental ice sheets. The Spanish are working on both of these problems, those related to separation of continental land mass history and those related to the ice-sheet history and climatic changes in Antarctica. These ice sheet studies are extremely relevant because by understanding the response of the ice sheets to past climate variability, it will help us understand what may happen with the ice sheet melting and sea level changes related to the present global warming.

The Antarctic ice sheet history in the Wilkes Land area of Antarctica has been the second phase of research that I have worked on with the Spanish scientists (see additional discussion in Chap. 5 Sect. 5.5). My wife, Dr. Carlota Escutia, studied seismic profiles penetrating deep below the seafloor to see the changes related to the ice-sheet history on the continental margin of Wilkes Land. She found that as the ice sheets were initially forming on Antarctica, large submarine landslides deposited sediment on the deep seafloor (Escutia et al. 2007). As the climate continued to become colder and the Antarctic Ocean circulation developed, turbidity currents and strong bottom currents deposited sand layers over the chaotic landslide debris [58–59]. Then while Carlota was chief scientist for the IODP Ocean Drilling leg 318, they drilled through these deposits and proved that her interpretation of the seismic profiles, ages of the deposits and interpretation of glacial history was correct [60].

As mentioned previously, the drill cores showed that 55 million years ago there was pollen from tropical African baobab and other palm trees [61]. This proved that at that time, Antarctica was a tropical greenhouse world. Then through the millions of years, the climate continued to cool, and ice sheets thousands of feet thick eventually covered Antarctica in the present icehouse world. This points out one of the false claims of climate change deniers when they say we are no different now than in the past, because climates were much warmer. The problem with this argument is that these changes took place over tens of millions of years when the world's biota and sea level could gradually adjust to the changing climate over millions of years.

Now in 150 years we have changed the CO_2 content to what it was 3 to 5 million years ago, when scientists have 95% confidence that sea level was about 22 m or over 70 feet higher [62]. In 3 to 5 million old Pliocene rock cliffs along the Ebro River valley in Spain I have observed sea urchins bored into the rocks at this height above sea level. Fortunately, our atmospheres present rapid change in CO_2 and global warming has thousands of years lag time before the full affects of sea level rise take place. Thus we need to understand the details of past global climate change to plan for the global warming that is rapidly happening.

My work assisting the Spanish scientists on their Antarctic projects has continued up until the present. I have been assisting studies of the deposits from the IODP 318 expedition to help to interpret the changes from the chaotic landslide debris to the combination of turbidity currents through channels and interaction with the strong Antarctic Bottom Water currents [63]. In 2017 I worked with the Spanish and Argentinean scientists to compare the seafloor deposits off Antarctica with the rocks at the tip of South America. We wanted to determine if the history of deposits in the rocks showed the same history as the seafloor deposits that we studied off Antarctica [58–59]. In the rocks, we needed to determine the interplay between bottom and turbidity current deposits. The past history of bottom currents is important to help us understand what may happen with glacial melting and ocean circulation related to the present climate change of global warming.

Recently I also have worked with other European scientists from the University of Ghent, Belgium and my wife's student Dr. Dimitris Evangelinos. We have been studying the glacial climate-related changes in the deposits of Lake Baikal Russia (Fig. 4.16). We observe that during the glacial periods of the past 2 million years, large turbidity currents occurred, which resulted in aprons and deepwater fans that deposited across the deep lake floor basins (Fig. 4.17) [64]. In contrast as the climate warmed between glacial periods around Lake Baikal, the glaciers melted, the forest growth returned, and the channels decreased in size and receded until the fan growth stopped.

These results showed that the development of the aprons and deepwater fans on the Lake Baikal floor was completely controlled by climate change. Previously scientists working on marine submarine fan systems related their development mainly to the lowering of sea level during glacial periods. These studies from Lake Baikal and the fact that during the Pleistocene ice ages of the past 2 million years, the number of submarine fans doubled in the ocean basins [34], both emphasize the importance of global climate change for the development of the largest sedimentary systems on our planet.

Also during my retirement, I have worked with scientists from the University of Texas at Austin and at Arlington to study the mini-basins on the continental slope and submarine fans on the deep-sea floor of the northern Gulf of Mexico (GOM) (Figs. 4.14, 4.15). We did these resource studies for a group of oil companies because these present-day systems could be used as analogues to better interpret similar ancient subsurface systems that are large petroleum producers. Our main research has taken place in offices and laboratories, where we compiled the data from our 1990 and 1997 USGS cruises, plus we added in all of the University of Texas data in our GOM study region. Our research thus includes the continental slope work on mini-basins, and comparison of three different types of submarine fans (Mississippi, Bryant and Rio Grande Fans) in the deep sea of the northern GOM (Fig. 4.14) [65, 33].

We studied mini-basins related to the Bryant Canyon system, which extended from an ancestral Mississippi Delta down to the deep seafloor (Fig. 4.15). This canyon connected through 15 mini-basins, and then fed the Bryant submarine fan [37]. Because the ancestral Mississippi River provided a huge sediment supply, as the sediment flows came down the canyon they would rapidly fill one of the mini-basins and then spell into the next basin fill it and so on until the entire chain of 15 basins was filled. Then the canyon had a continuous pathway all the way down the continental slope and could feed directly into the Bryant Fan on the deep seafloor.

We found that the deposits filling each mini-basin and the connecting Bryant Canyon segments between the basins had the same pattern. Layers of submarine landslide chaotic deposits alternated with ponded turbidite sand layers and the latter provide the good petroleum reservoirs in the ancient subsurface mini-basins [33]. Above this alternating sequence of landslide and turbidite units, a layer of multiple bypass channels was found because these provided the pathway to transport sediment over a filled basin into the next unfilled basin. After a mini-basin was filled with sediment, the weight of the sediment would cause it to sink into the underlying salt and push up the salt into ridges around the basin [37]. Consequently, high relief walls covered by thick muds surrounded these small mini-basins. As a result, multiple submarine landslides came off the steep mini-basin walls and covered each basin with a thick layer of submarine landslide deposits.

An analysis of the GOM submarine fan turbidite systems has continued to the present (Fig. 4.14, 4.15). Our project analyzed the Bryant Fan and the Rio Grande Fan and compared them to the present day Mississippi Fan [65]. The Rio Grande Fan (100 km/62 mile length) has multiple braided channel systems fed by many slope canyons. This fan provides an analogue for subsurface petroleum systems in the northwest GOM. The Bryant Fan (200 km/125 mile length) has single large channel levee systems that feed single sand lobes at the distal end of the fan (Fig. 4.15). It provides an analogue for the 20 to 30-million-year old ancient petroleum-rich fan deposits in the Mississippi Canyon area of the GOM.

The Mississippi Fan (600 km/372 mile length) has large proximal channel levee systems that continue splaying into complex smaller and smaller channels across the fan (Fig. 4.14 see red box). These channels transport both submarine landslide debris flow mud and sand [36]. Also occasional huge submarine landslides cover large parts of Mississippi Fan (Fig. 4.14 see orange area) [66]. The Mississippi Fan provides a modern analogue for the large petroleum-rich ancient submarine fans thousands of feet below the GOM seafloor, which have been discovered to be petroleum sources during the past decade.

The result from studying the many different types of mini-basin and submarine fan turbidite systems is three large several-hundred page atlases. These atlases provide interpretations of the seismic profiles, description of the sediment from cores and analyses of the sediment ages from the microscopic fossils [67]. The atlas data from modern turbidite systems provide analogues that the oil companies can use to help explore for and understand the petroleum reservoirs in similar ancient subsurface deposits of the GOM [65, 33].

Since 2006 I also have worked on a similar project studying submarine fan turbidite systems at the University of Leeds in England. This University has another group of oil companies that sponsor academic research on deep-water sedimentary systems as it relates to petroleum systems. Again, I compiled all my worldwide experience on modern deep-water systems and summarized this information in a series of lectures that were given at the company sponsored meetings in the United States and Europe. In Europe we had field trips to visit rock sequences that had similar types of deposits to those we studied on the modern seafloor. These comparisons help companies to develop petroleum reservoirs and increase oil resources.

My research activities, with many international universities, government agencies and industry since 2000, have kept me current with my lifelong studies on geological resources, hazards, and environmental assessment. As a result, I have continued to learn about both natural and human-caused global change.

References

1. Nelson CH, Carlson PR, Bacon CR (1988) The Mt. Mazama climactic eruption and resulting convulsive sedimentation on the continent, ocean basin, and Crater Lake caldera floor. In: Clifton HE (ed) Sedimentologic consequences of convulsive geologic events, geological society of America special paper, 229, pp 37–56
2. Nelson CH, Bacon C (2004) Crater lake: presently Tranquil, USGS, selected issues in the USGS marine and coastal geology program, United States department of the interior, U.S. geological survey, p 3. craterlakeinstitute.com/geology/natural-history-geology-articles/geology-stuff-04/
3. Nelson CH, Bacon CR, Robinson SW, Adam DP, Bradbury JP, Barber JH Jr, Schwartz D, Vagenas G (1994) The volcanic, sedimentologic and paleolimnologic history of the Crater Lake caldera floor, Oregon: evidence for small caldera evolution. Geol Soc Am Bull 106:684–704
4. Nelson CH, Kulm LD, Carlson PR, Duncan JR (1968) Mazama ash in the northeastern Pacific. Science 161:47–49
5. Nelson CH, Escutia C, Karabanov EB, Colman SM (2000) Tectonic and sediment supply control of deep rift lake turbidite systems. Lake Baikal, Russia, Reply, Geology 28:190–191
6. Nelson CH, Chris Goldfinger, Joel E Johnson, Gita Dunhill (2000) Variation of modern turbidite systems along the subduction zone margin of Cascadia

basin and implications for turbidite reservoir beds. In: Weimer PW, Nelson CH et al. (eds.) Deep-water Reservoirs of the World, Gulf Coast Section Society of Economic Paleontologists and Mineralogists Foundation 20th Annual Research Conference, CD ROM:714-738

7. Nelson CH, Meyer AW, Thor D, Larsen M (1986) Crater Lake, Oregon: a restricted basin with base-of-slope aprons of nonchannelized turbidites. Geology 14:238–241

8. Nelson CH, Johnson KR (1987) Whales and walruses as tillers of the sea floor. Sci Am 256:112–117

9. Nelson CH, Rowland RW, Stoker S, Larsen BR (1981) Interplay of physical and biological sedimentary structures of the Bering Sea epicontinental shelf. In: Hood DW, Calder JA (eds.) The eastern bering sea shelf: its oceanography and resources, NOAA-OCSEAP Juneau Regional Office, Seattle, University of Washington Press, 2:1265–1296

10. Nelson CH (1982a) Holocene transgression in deltaic and non-deltaic areas of the Bering epicontinental shelf. In: Nelson CH, Nio SD (eds.) The Northeastern Bering Shelf: New Perspectives of Epicontinental Shelf Processes and Depositional Products. Geologie en Mijnbouw 61:5–18

11. Johnson KR, Nelson CH (1984) Side-Scan sonar assessment of gray whale feeding in the Bering Sea. Science 225:1150–1152

12. Nelson CH, Johnson KR, Barber JH Jr. (1987) Extensive whale and walrus feeding excavation on the Bering shelf, Alaska. J Sedim Petrol 57:419–430

13. Fay FH, Eberhardt LL, Kelly BP, Burns JJ, Quakenbush LT (1997) Status of the Pacific walrus population, 1950–1989. Mar Mam Sci 13:537–565

14. Netflex (2019) Our Planet, Episode 2 Frozen worlds. https://www.netflix.com/es-en/title/80049832

15. Harris M (2015) Waves of destruction. Sci Am 312:64–69

16. MacCracken JG, William S Beatty, Joel L Garlich-Miller, Michelle L Kissling, Jonathan A Snyder (2017) Final species status assessment for the Pacific Walrus (Odobenus rosmarus divergens), May 2017 (Version 1.0), U.S. Fish and Wildlife Service, Marine Mammals Management, 1011 E. Tudor, Rd.MS-341, Anchorage, AK 99503, p 295

17. Higdon JW, DB Stewart (2018) State of circumpolar walrus (Odobenus rosmarus) populations. Prepared by Higdon Wildlife Consulting and Arctic Biological Consultants, Winnipeg, for the World Wildlife Federation Arctic Programme, Ottawa, ON. p 100

18. Nelson CH, Maldonado A (1990) Factors controlling Late Cenozoic Ebro margin growth from the Ebro delta to the western Mediterranean deep sea. In: Nelson CH, Maldonado A (eds.) The Ebro continental Margin, Northwestern Mediterranean Sea, Marine Geology Special Issue 95: 419–440

19. Hsu¨ KJ, Ryan WBF, Cita MB (1973) Late Miocene desiccation of the Mediterranean. Nature 242:240–244

20. Escutia C, Maldonado A (1992) Paleoceanographic implications of the Messinian surface in the Valencia Trough, northwestern Mediterranean Sea. Tectonophysics 203:263–284

21. Garcia-Castellanos D, Estrada F, Jiménez-Munt I, Gorini C, Fernàndez M, Vergés J, De Vicente R (2009) Catastrophic flood of the Mediterranean after the Messinian salinity crisis. Nature 462:778–781. https://doi.org/10.1038/nature08555

22. Ryan WBF (2009) Decoding the Mediterranean salinity crisis Sedimentology 56: 95–136. DOI:https://doi.org/10.1111/j.1365-3091.2008.01031.x

23. Dan G, Nabil S, Savoye B (2007) The 1979 Nice harbour catastrophe revisited: Trigger mechanism inferred from geotechnical measurements and numerical modelling. Mar Geol 245:40–64. http://dx.doi.org/10.10

24. Nelson CH (1990) Estimated Post-Messinian sediment supply and deposition rates of the Spanish Ebro margin. In: Nelson CH, Maldonado A (eds.) Marine Geology of the Ebro Continental Margin, Northwestern Mediterranean Sea, Marine Geology Special Issue 95:395–418

25. Steffen W, Broadgate W, Deutsch L, Gaffney O, Ludwig C (2015) The trajectory of the Anthropocene: the great acceleration. Anthrop Rev 2:81–98

26. Syvitski JPM, Saito Y (2007) Morphodynamics of deltas under the influence of humans, Global Planet. Changes 57:261–282

27. Stanley DJ, Warne AG (1993) Nile Delta: recent geological evolution and human impacts. Science 260:628–634

28. Shultz D (2020) Dust storms associated with increase in critical care visits. Eos, 101. https://doi.org/10.1029/2020EO147660

29. Nelson CH, Larsen BR, Jenne EA, Sorg DH (1977) Mercury dispersal from lode sources in the Kuskokwim River drainage, Alaska. Science 1198:820–824

30. van Geen A, Adkins JF, Boyle EA, Nelson CH, Palanques A (1997) A 120 year record of metal contamination on an unprecedented scale from mining of the Iberian Pyrite Belt. Geology 25:291–294

31. Maldonado A, Nelson CH (1999) Interaction of tectonic and depositional process that control the evolution of the Iberian Gulf of Cadiz. In: Mandonado A, Nelson CH (eds.) Marine Geology of the Gulf of Cadiz. Mar. Geol. Special Issue 155:217–242

32. Nelson CH, Baraza J, Maldonado A, Rodero J, Escutia C, Barber Jr JH (1999) Influence of the Atlantic inflow and Mediterranean outflow currents on Late Quaternary sedimentary facies of the Gulf of Cadiz continental margin, In: Mandonado A, Nelson CH (eds.) Marine Geology of the Gulf of Cadiz. Mar. Geol. Special Issue 155:99–129

33. Nelson CH, Damuth JE, Olson HC (2018) Late Pleistocene Bryant Canyon turbidite system: Implications for Gulf of Mexico minibasin petroleum systems, Interpretation 6:SD89–SD114. http://dx.doi.org/https://doi.org/10.1190/INT-2017-0150.1

34. Nelson CH, Nilsen TH (1984) Modern and ancient deep-sea fan sedimentation, Society of economic geology and paleontology short course 14, Tulsa, OK., p 403

35. Twichell DC, Schwab WC, Nelson CH, Kenyon NH, Lee HJ (1992) Characteristics of a sandy depositional lobe on the outer Mississippi fan from Seamarc 1A Sidescan Sonar Images. Geology 20:689–692

36. Nelson CH, Twichell CC, Schwab WC, Lee HJ, Kenyon NH (1992) Late Pleistocene Turbidite Sand Beds and Chaotic Silt Beds in the Channelized Distal Outer Fan Lobes of Mississippi Fan, Geology 20:693–696

37. Twichell DC, Nelson CH, Damuth JE (2000) Late-stage development of the Bryant Canyon turbidite pathway on the Louisiana continental slope: In: Weimer P et al. (eds.) Deep-Water Reservoirs of the World, 20th Annual GCSSEPM Foundation Bob F. Perkins Research Conference, CD-ROM, pp 1032–1044

38. Nelson CH, Karabanov EB, Colman SM (1995) Late Quaternary Lake Baikal turbidite systems, Russia. In: Pickering KT, Lucchi FR, Smith Ru, Hiscott RN, Kenyon N (eds.) An Atlas of Deep-Water Environments, Chapman and Hall, London, pp 29–33

39. Nelson CH, Karabanov EB, Colman SM, Escutia C (1999) Tectonic and sediment supply control of deep rift lake turbidite systems. Lake Baikal, Russia, Geology 27:163–166

40. Nelson CH, Savoye B, Fremer I, Rehault JP, Escutia C (1996) Interfingering Sand-Rich Aprons and Var Fan Lobe Deposits off Corsica: Analog for Thick and Laterally Extensive Turbidite Petroleum Reservoirs, 1996 American Association of Petroleum Geologists Annual Convention, San Diego, Official Program 5:A104

41. Rothwell RG, Reeder MS, Anastasakis G, Stow DAV, Thomson J, Kahlera G (2000) Low sea-level stand emplacement of megaturbidite in the western and eastern Mediterranean Sea. Sed Geol 135:75–88

42. Polonia A, Bonatti E, Camerlenghi A, Lucchi RG, Panieri G, Gasperini L (2013) Mediterranean megaturbidite triggered by the AD 365 Crete earthquake and tsunami. Nat Sci Rep 3:1–12. https://doi.org/10.1038/srep01285

43. Polonia A, Nelson CH, Romano S, Vaiani SC, Colizza E, Gasparotto G, Gasperini L (2017) A depositional model for seismo-turbidites in confined basins based on Ionian Sea deposits. Marine Geol 384:177–198. ISSN 0025-3227. https://doi.org/10.1016/j.margeo.2016.05.010

44. Lansing A (1999) 2nd ed. Endurance: Shackleton's Incredible Voyage, Carroll & Graf Publishers. ISBN 0–7867–0621-X

45. Gutierrez-Pastor J, Nelson CH, Goldfinger C, Johnson JE, Escutia C, Eriksson A, Morey A, The Shipboard Scientific Party (2009) Earthquake control of Holocene turbidite frequency confirmed by hemipelagic sedimentation chronology on the Cascadia and northern California active continental margins. In: Kneller B, McCaffrey W, Martinsen OJ (eds.) SEPM, Special Publication 92:179–197

46. Goldfinger C, Ikeda Y, Yeats RS, Ren J (2013) Superquakes and supercycles. Seismol Res Lett 84:24–32. https://doi.org/10.1785/0220110135
47. Goldfinger C, Nelson CH, Morey AE, Gutierrez-Pastor J, Johnson JE, Karabanov E, Chaytor J, Dunhill G, Ericsson A (2012) Rupture lengths and temporal history of Cascadia great earthquakes based on turbidite stratigraphy, USGS Professional Paper 1661-F. In Kayen R (ed) Earthquake hazards of the Pacific Northwest Coastal and Marine regions, p 192
48. Adams J (1990) Paleoseismicity of the Cascadia subduction zone—Evidence from turbidites off the Oregon-Washington margin. Tectonics 9:569–583
49. Atwater BF (1987) Evidence for great Holocene earthquakes along the outer coast of Washington State. Science 236:942–944
50. Nelson CH (1976) Late Pleistocene and Holocene depositional trends, processes and history of Astoria deep-sea fan, northeast Pacific. Marine Geol 20:129–173
51. Nelson CH, Goldfinger C, Gutierrez-Pastor J (2012) Great earthquakes along the western United States continental margin: implications for hazards, stratigraphy and turbidite lithology. In: Pantosti D, Gràcia E, Lamarche G, Nelson CH (eds.) Special Issue on Marine and Lake Paleoseismology, Nat. Hazards Earth Syst Sci 12:3191–3208
52. Goldfinger C, Grijalva K, Burgmann K, Morey A, Johnson JE, Nelson CH, Gutiérrez-Pastor J, Karabanov E, Patton J, Gracia E (2008) Late Holocene rupture of the northern San Andreas Fault and possible stress linkage to the Cascadia Subduction Zone. Bull Seismol Soc Am 98:861–889
53. Patton JR, Goldfinger C, Morey AE, Romsos C, Black BY, Djadjadihardja A (2013) Seismoturbidite record as preserved at core sites at the Cascadia and Sumatra–Andaman subduction zones. Nat Hazards Earth Syst Sci 13:833–867. www.nat-hazards-earth-systsci.net/13/833/2013/, DOI:https://doi.org/10.5194/nhess-13-833-2013
54. Strasser M, Moneke K, Schnellmann M, Anselmetti FS (2013) Lake sediments as natural seismographs: a compiled record of Late Quaternary earthquakes in Central Switzerland and its implication for Alpine deformation. Sedimentology 60:319–341. https://doi.org/10.1111/sed.12003
55. Nilsen TH (2000) The Hilt Bed, an Upper Cretaceous compound basin–plain seismo-turbidite in the Hornbrook Forearc Basin of southern Oregon and northern California, USA. Sed Geol 135:51–63
56. Maldonado A, Barnolas A, Bohoyo F, Escutia C, Galindo-Zalzivar J, Hernandez-Molina FJ, Jabaloy A, Lobo F, Nelson CH, Rodríguez-Fernandez J, Somoza L, Suriñach E, Vazquez JT (2005) Miocene to Recent Contourite Drifts Development in the Northern Weddell Sea (Antarctica): Special Issue on Global and Planetary Change 45:99–129
57. Maldonado A, Barnolas A, Bohoyo F, Escutia C, Galindo-Zaldívar J, Hernández-Molina J, Jabaloy A, Lobo FJ, Nelson CH, Rodríguez-Fernández J, Somoza L, Suriñach E, Vázquez JT (2006) Seismic stratigraphy of the Scotia Sea and Weddell Sea deposits, Miocene to Recent sedimentary deposits in the

central Scotia Sea and northern Weddell Sea (Antarctica): influence of bottom flows. In: Fütterer DK, Damaske D, Kleinschmidt G, Miller H, Tessensohn F (eds) Antarctica: contributions to global earth sciences. Springer-Verlag, Berlin Heidelberg New York, pp 441–446

58. Escutia C, Eittreim SL, Cooper AK, Nelson CH (2000) Morphology and acoustic character of the Antarctic Wilkes Land turbidite systems: ice-sheet-sourced versus river-sourced fans. J Sediment Res 70:84–93

59. Escutia C, Nelson CH, Acton GD, Eittreim SL, Cooper AK, Warnke DA, Jaramillo JM (2002) Current Controlled Deposition on the Wilkes Land Continental Rise, Antarctica. In: Stow DAV, Faugeres J-C, Howe JC, Pudsey C, Viana A (eds.) Deep-Water Contourite Systems: Modern Drifts and Ancient Series, Seismic and Sedimentary Characteristics, Geological Society London Memoirs 22:373–384

60. Escutia C, Brinkhuis H, Klaus A, the Expedition 318 Scientists (2011) Expedition 318, Site U1356, Proc. IODP 318. DOI:https://doi.org/10.2204/iodp.proc.318.104.2011

61. Pross J, Contreras L, Bijl PK, Greenwood DR, Bohaty SM, Schouten S, Bendle JA, Röhl U, Tauxe L, Raine JI, Huck CE, van de Flierdt T, Jamieson SSR, Stickley CE, van de Schootbrugge B, Escutia C, Brinkhuis H, IODP Expedition 318 Scientists (2012) Persistent near-tropical warmth on the Antarctic continent during the early Eocene epoch. Nature 488:73–77. DOI:https://doi.org/10.1038/nature11300

62. Miller KG, 9 others (2012) High tide of the warm Pliocene: implications of global sea level for Antarctic deglaciation, Geology 40:407–410. https://doi.org/10.1130/G32869.1

63. Salabarnada A, Escutia C, Rohel U, Nelson CH, McKay R, Jiménez-Espejo FJ, Bijl PK, Hartman J, Strother S, Salzmann U, Evangelinos D, López-Quirós A, Flores JA, Sangiorgi F, Ikehara M, Brinkhuis H (2018) Paleoceanography and ice sheet variability offshore Wilkes Land, Antarctrica—Part 1: Insights from late Oligocene astronomically paced contourite sedimentation. Climate of the past 14:991–1014

64. Evangelinos D, Nelson CH, Escutia C, De Batist M, Khlystov O (2017) Late quaternary climatic control of Lake baikal turbidite systems: implications for turbidite systems worldwide. Geology 45:179–182. https://doi.org/10.1130/G38163.1

65. Nelson CH, Damuth JE, Olson H, Escutia C (2010) Factors controlling modern submarine fan architecture and implications for paleogene to Miocene petroleum plays, Gulf of Mexico, AAPG Search and Discovery.com, number 50287, p 35

66. Twichel D, Nelson CH, Kenyon N, Schwab W (2009) The influence of external processes on the latest Pleistocene and Holocene evolution of the Mississippi Fan. In: Kneller B, McCaffrey W, Martinsen OJ (eds.) SEPM Special Publication 92:145–157

67. Olson HC, Damuth JE, Nelson CH (2016) Latest quaternary sedimentation in the northern Gulf of Mexico intraslope basin province: II—Stratigraphic analysis and relationship to glacioeustatic climate change, Interpretation (SEG/AAPG), v. 4, pp 1–15. http://dx.doi.org/https://doi.org/10.1190/INT-2015-0111.1

68. Nelson CH, Escutia C, Goldfinger C, Twichell DC, Damuth JE (2011) Case studies comparing mass-transport and turbidite-system deposits in different continental margin settings and time periods. In: Shipp C, Weimer P, Posimentier H (eds.) SEPM Special Publication 95:39–66

69. Liu JY, Bryant WR (2000) Seafloor relief of northern Gulf of Mexico deep water, Texas Sea Grant College Program

70. Suter JR, Berryhill HL (1985) Late Quaternary shelf margin deltas, northwest Gulf of Mexico, AAPG Bulletin 69:77–81

71. Jorry SJ, Jegou I, Emmanuel L, Jacinto RS, Sovoye B (2011) Turbiditic levee deposition in response to climate changes: The Var Sedimentary Ridge (Ligurian Sea). Marine Geol 279:148.161

72. Satake K, Shimazaki K, Tsuji Y, Ueda K (1996) Time and size of a giant earthquake in Cascadia inferred from Japanese tsunami records of January, 1700. Nature 379:246–249

73. Newhall C, Self S, and Robock A (2018) Anticipating future Volcanic Explosivity Index (VEI) 7 eruptions and their chilling impacts. Geosphere 14:572–603

74. Nelson CH, Maldonad A (1988) Factors controlling depositional patterns of ebro turbidite systems, Mediterranean Sea. AAPG Bulletin 72

75. Nelson CH, Lamothe PJ (1993) Heavy Metal Anomalies in the Tinto and Odiel River and Estuary system, In: Kuwabara, JS (ed.) Trace Contaminants and Nutrients in Estuaries, Estuaries 16:496–511

Part II

Earth in Peril

5

Climate Change Patterns

5.1 Climate and Global Warming Background

I will begin my discussion on human-caused global change by describing climate change and global warming, because this change is most pervasive throughout the Earth and this topic is the most common in the media. In Part I of my geologist's journey, I covered both human-caused and natural global changes, however the rest of the book will focus on human-caused changes.

Before beginning my description of global warming patterns, I should point out that I am not a climate scientist. However as described in the preface, I have undergraduate, MS and Ph.D. graduate school training and research experience in biology, geology, limnology (lake studies) and oceanography. My global experience, especially in limnology and oceanography, does provide a background to help me evaluate the scientific evidence that there is human-caused global warming. An interesting anecdote is that many of the leading scientists investigating global warming have oceanography backgrounds [23].

Following are some terms to keep in mind during my discussion about global warming:

Human-caused global change refers to all the human-caused changes to our Earth's air, water and soil. One of the most important take home messages of this book is that human-caused climate change and global warming are just one of the many global changes threatening the human species.

© The Author(s), under exclusive license to Springer Nature
Switzerland AG 2021
C. H. Nelson, *Witness To A Changing Earth*,
https://doi.org/10.1007/978-3-030-71811-4_5

Weather is the day-to-day change in temperature, cloud cover, precipitation and wind. People often confuse the terms weather and climate, such as President Trump inferring that the 2017 cold New Years Eve temperature in New York City indicated that global warming does not exist. The cold 2017 New Year's temperature was part of the day-to-day weather.

Climate is the long-term weather of many years.

Climate change is the long-term change in weather patterns that takes place over many years. Climate change can result from natural causes and is called ***natural climate variability***, which has controlled climate change throughout geologic time, until the industrial revolution of the last 150 years. Climate change also can result from human activities, especially the burning of fossil fuels, and this should be labeled ***human-caused climate change***. The term climate change now is often substituted as shorthand for human-caused climate change because the topic is so common in the media and political discussions. For example, the term climate change denier or skeptic is used to describe people that do not believe that humans have caused the recent rapid rise of global temperatures. However the correct term should be human-caused climate change denier/skeptic, although this terminology is never used. Because of the common shorthand usage of the term climate change for human-caused climate change, I often use it interchangeably with global warming in this book.

Global warming refers to the earth's warming climate that 100% of peer-reviewed articles from climate scientists attribute to human activities, primarily fossil fuel burning [41].

As described in the following sections of this chapter, there are an overwhelming number of scientific facts showing that the climate is warming fast compared to past geologic time. All the evidence indicates that humans cause recent climate warming (see summary in Sect. 5.9 of this chapter). This rapid global warming results in environmental change that is affecting all life on the planet with a huge cost in lost lives and economic damage. There also is unequivocal evidence that humans are causing many other global changes that are detrimental to life on Earth, whether or not climate change results from humans.

Part of the problem that can lead to global warming denial is that climate change is investigated by natural science methods. Natural science is not a hard science like math and physics where you can write a mathematical formula like $E = mc^2$ for relatively. The natural world has myriads of complexities and one observation may indicate one trend and another may indicate an opposite trend. For example, overall the temperatures near the earths land surface are warming significantly. However, there are some periods

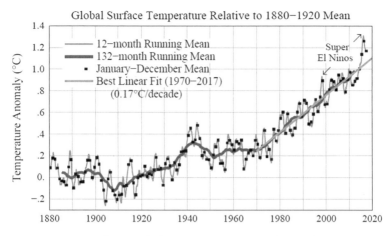

Fig. 5.1 Increasing surface global temperature anomalies relative to 1880–1920 based on GISTEMP data, which employs GHCN.v3 for meteorological stations, NOAA ERSST.v5 for sea surface temperature, and Antarctic research station data1. Figure source is Hansen et al. [15]

where temperatures have not been consistently warming such as 1940s to 1970s (Fig. 5.1) [15].Deniers/skeptics can look at a time period when the temperature is not rising and say there is no global warming. In contrast to the more variable land surface temperatures, the upper oceans are warming with little fluctuation, which show continuous global warming since the 1950s (Fig. 5.2) [5]. Thus for natural sciences like the earth sciences, the best scientific evidence for global warming is many scientific observations that show the same dominant pattern of human-caused climate change. Following are some of the main patterns of the earth's climate change that scientists use to provide evidence for human-caused global warming.

5.2 Increased Greenhouse Gases

There have always been fluctuations of carbon dioxide (CO_2) in the atmosphere (Figs. 5.2, 5.3, 5.4g and 5.5). However there is an obvious pattern showing that the yearly input of CO_2 has increased 350% while the population has increased 250% since 1960 (Figs. 5.2 and 5.4g, i). The most important human activity, that increases CO_2, is the burning of fossil fuel. This combustion releases CO_2 and soot into the atmosphere. In addition, humans cause the release of methane (CH_4) during the production of fossil fuels and livestock (Fig. 5.6) (https://en.wikipedia.org/wiki/Methane_emissions). There is an additive effect of warming climate, which causes melting

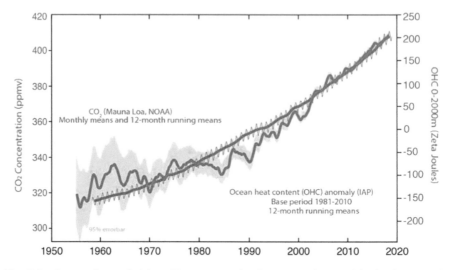

Fig. 5.2 Comparison of rising CO_2 content in the atmosphere with the increase in ocean heat content in the upper 2000 m (6,500 feet) of the ocean since 1960. Figure source is Cheng et al. [63]

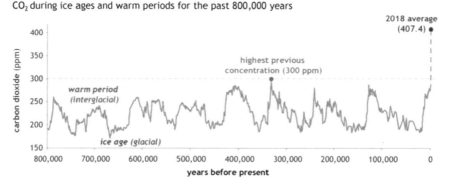

Fig. 5.3 Atmospheric CO_2 for the past 800,000 years. Note the rapid increase since the industrial revolution. Figure source is NOAA Climate.gov National Center for Environmental Information

of frozen permafrost and frozen CH_4 below the seafloor to release CH_4 into the atmosphere [32].

Water vapor in the atmosphere is the major greenhouse gas. Increased CO_2 and CH_4 in the atmosphere drive exponential increases in water vapor, which then traps more of earth's reflected heat. The human-caused introduction of soot also helps trap heat and deforestation reduces the amount of CO_2 that is absorbed from the atmosphere (Fig. 5.4e). All of these human activities from

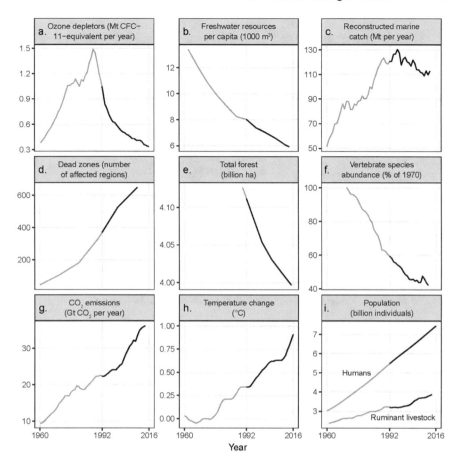

Fig. 5.4 Trends over time for environmental issues identified in the 1992 scientist's warning to humanity. The years before and after the 1992 scientists' warning are shown as gray and black lines, respectively. **a** Reduced ozone depletors (emissions of halogen source gases, which deplete stratospheric ozone, assuming a constant natural emission rate of 0.11 Mt CFC-11-equivalent per year). **b** Reduced freshwater per capita (e.g. Figs. 6.4, 6.5) **c** Reduced marine catch since the mid-1990s. **d** Increase of ocean dead zones. **e** Reduced total forest area. **f** Reduced vertebrate species (-58%, with freshwater, marine, and terrestrial populations declining by 81, 36, and 35%, respectively. **g** Yearly increase in CO_2 emissions. **h** Temperature Increase shown in five-year means. **i** Increase in human population and ruminant livestock (domestic cattle, sheep, goats, and buffaloes). Note that y-axes do not start at zero, and it is important to inspect the data range when interpreting each graph. Percentage change, since 1992, for the variables in each panel are as follows: (a) -68.1%; (b) -26.1%; (c) -6.4%; (d) +75.3%; (e) -2.8%; (f) -28.9%; (g) +62.1%; (h) +167.6%; and (i) humans: +35.5%, ruminant livestock: +20.5%. Sources for Fig. 5.4 are in file S1 of Ripple et al. [62]

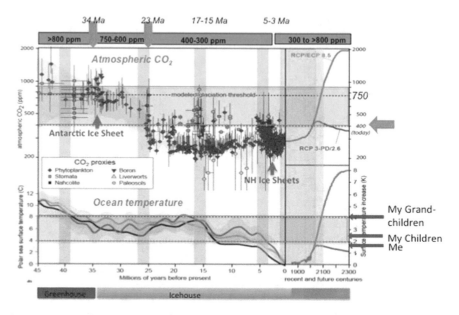

Fig. 5.5 Forty five million year history of atmospheric CO_2 content, Antarctic ice sheet development from greenhouse to icehouse world, Northern Hemisphere (NH) formation of ice sheets ([19] Chap. 6 WG1), and polar sea surface temperatures [6]. In the right-hand column from year 1900 of our era to year 2300 yr (projected) are the changes **a** if the average global temperature increase is near 1 °C (2.12 °F) and CO_2 does not increase beyond the present day content of about 400 ppm (blue line and large orange arrow on right of diagram) or **b** if temperature increases as much as 8 °C (17 °F) (lower orange line) and CO_2 content increases to as much as 2000 ppm (upper orange line). If CO_2 reaches 750 ppm, it is projected that there will be a loss of Antarctic ice sheets and sea level would rise tens of meters [11]. Most important, the present atmospheric CO_2 amount of 400+ ppm (see Figs. 5.2, 5.3) and Arctic sea surface temperature (4 °C, 7.2 °F) (see Figs. 4.6 and 4.9) [40] already are equal to those of the warmer Pliocene world of 3–5 million years ago. Figure source Cramer et al. [6] IPCC [19] (modified from Carlota Escutia, Instituto Andaluz de Ciencias de la Tierra, 18100, written communication 2020)

burning fossil fuels and cutting forests contribute to warming of the atmosphere. The warmer atmosphere then absorbs more water vapor, and thus increased CO_2 and CH_4 are triggers or amplifiers for water vapor. Consequently humans need to control our input of CO_2 and CH_4 where their increase correlates with atmospheric and oceanic temperature rise and also affects the amount of water vapor in the atmosphere (Figs. 5.1, 5.2, 5.3 and 5.4g, h). This correlation of more CO_2 in the atmosphere and increasing average global ocean temperature has been observed for tens of millions of years (Fig. 5.5). However, the present pattern of the recent rapid increase, of CO_2, CH_4 and temperature has occurred during the past 170 years of the

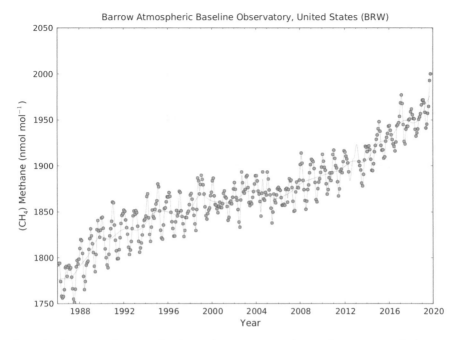

Fig. 5.6 Content of atmospheric methane measured at Point Barrow Alaska USA since 1986. Note that the methane content has been rising since that time and the content is accelerating since 2008. The recent increase may be related to (1) amphlified global warming in the Arctic (e.g. Fig. 5.6) causing release of methane from melting permafrost and frozen methane under the sea, and (2) the increased petroleum development such as fracking. Figure source is https://en.wikipedia.org/wiki/Arctic_methane_emissions

post-industrial revolution during increased fossil fuel burning and has been accelerating in the 21st century (Figs. 5.1, 5.2, 5.3, 5.4g, h and 5.5).

Climate change deniers/skeptics mantra is the earth's climate has been much warmer in the past, which is definitely true. Certainly in the past, natural CO_2 and average global temperatures have been equal or much higher than at present (Fig. 5.5). My wife and her scientific team found pollen related to the tropical African baobab tree in cores drilled deep below the seafloor in 55 million year old sediment off Antarctica [42]. However the important factor is that natural climate change from the Antarctic tropical greenhouse world to the icehouse world with an Antarctic ice sheet took place over a long time period from 45 to 34 million years ago (Fig. 5.5). The present rapid increase in CO_2 of the past century has resulted in a human-caused CO_2 amount that has not been seen on earth since 3–3.5 million years ago (Figs. 5.1, 5.2, 5.3, 5.4g and 5.5). This extremely rapid unnatural increase of CO_2 and average global temperatures during the past century has not been observed in previous geologic history (e.g. Trapati et al. [54]). The only similar rapid natural climate change in geologic history occurred when a large meteorite or asteroid hit the earth 66 million years ago and 70% of the Earth's species, including dinosaurs, became extinct [1].

I said mainly CO_2 emissions are responsible for the increasingly hot weather, however CH_4 is another important greenhouse gas that has been increasing in the atmosphere even faster than CO_2 (Fig. 5.6). In North America since the industrial revolution, 60% of CH_4 release has resulted from human activity and CH_4 has increased 160% compared to 42% for CO_2 (https://en.wikipedia.org/wiki/Methane_emissions; [32]). The methane amount in the atmosphere has increased 2.5 times since pre-industrial times and like CO_2 is the highest it has been in the past 800,000 years (Figs. 5.3 and 5.6).

Even though CH_4 remains much less time in the atmosphere than CO_2, its increase is a major concern because over 100 years it is 34 times more potent as a greenhouse gas than CO_2 and is responsible for 25% of the present global warming temperatures (https://en.wikipedia.org/wiki/Methane_emis sions; [32]). Although fossil fuel activities and some other human activities like raising livestock provide about one third each of the unnatural sources of CH_4, 40% of the total input results from biologic activity such as in wetlands. Unnatural wetland activity is increasing rapidly because it is linked to global warming and melting permafrost. Likewise warming temperatures result in unnatural thawing of frozen methane (CH_4) that is released into the atmosphere. This will be described in detail in my discussion in Sect. 5.5.

There are solutions to reduce CH_4 input into the atmosphere. The most important is to lower CO_2 and slow global warming so that there is less melting of the vast areas of frozen methane and permafrost, which will reduce biological input of CH_4. Eating less meat and cutting livestock production also can reduce biological input of CH_4 (Fig. 5.4i). Reducing meat in the diet is a win win solution because less livestock is good for both human and the Earth's health. Another gain for humans and our planet results from the USA regulations to cut CH_4 emissions from petroleum production by 70%. Unfortunately, similar to many environmentally sound regulations, from 2017 to 2020 the EPA (Environmental Protection Agency) in the USA eliminated the regulation to reduce CH_4 release by the fossil fuel industry.

5.3   Acid Ocean

Ocean waters are absorbing more of the increased CO_2 from the atmosphere and this may be as grave a problem as global warming from CO_2 (Fig. 5.2). When the increased CO_2 is absorbed into the ocean water, it creates more carbonic acid and the entire ocean becomes more acid. This is affecting the base of the oceanic food chain because the thin carbonate shells of microscopic plants and animals can dissolve from the increased acid. These plants, animals and small larval stages provide food for other larger ocean species.

In the greenhouse world of 45 million years ago (Fig. 5.5), when the ocean surface was several degrees warmer and more acidified than now, there was significant extinction of bottom dwelling species (https://en.wikipedia.org/wiki/Ocean_acidification). Climate change deniers can say that the planet survived this, so why worry. The problem is that the present rate of ocean acidification is occurring 100 times faster than in the past 20 million years. The rate of ocean acidification has doubled in the past 40 years and is on a path to reach amounts not seen in the last 65 million years. If we continue emitting CO_2 at the same rate, by 2100 ocean acidity will increase about 150%, a rate that has not been experienced for at least 400,000 years (http://www.oceanacidification.org.uk/). In the past species had millions of years to evolve and adapt to changing ocean acid conditions, However, with the present rapid rates of acidification, adaption might not be possible for many species and ocean food sources are in peril. For example, new studies of Dungeness crab in the Pacific Ocean show some dilution of the carbonate shells of crab larvae, which potentially affects sensory and behavorial patterns, and can lead to reduction of this important food supply [2].

The sea butterfly plankton is another example of the threat of ocean acidification reported by [28]. A cubic mile of the surface of the Southern Ocean contains 27 trillion sea butterflies. This plankton is one of the primary foundations of the food chain that eventually supports the larger species of fish and mammals. The butterflies have a fragile thin carbonate shell a centimeter (half-inch) across but only as thick as a human hair. Experiments with presently acidified seawater show that after 6 weeks butterfly shells are destroyed. This will be a huge loss for the Antarctic and other ocean area food webs and also for removing CO_2 from the atmosphere. Because of their great abundance, the butterfly shells remove significant amounts of CO_2 from the atmosphere to build their shells, which eventually sink, deposit on the seafloor, and permanently reduce CO_2 from the atmosphere.

5.4   Rising Ocean Temperature and Sea Level

The patterns of rising temperature and sea level of the ocean provide the best evidence for global warming, because the ocean is 70% of our Earth's environment and requires a huge input of melting ice to raise sea level and significant temperature increase to heat it (Figs. 5.2, 5.7 and 5.8). The ocean absorbs 93% of the planets increased heat content, whereas the atmosphere only absorbs 2% of the increased heat from human activities [5]. However, if the atmosphere had absorbed all the increased heat, the global temperature already would have raised 60 °C (108 °F). This shows the importance of the ocean for moderating the temperature rise from global warming. The massive amount of new measured average ocean temperatures to depths of 2000 m of water depth (about 6,000 feet) verifies climate models. Unfortunately these data indicate that ocean warming is 40% faster than thought and will increase 6 times by 2081 to 2100 (Fig. 5.2) [5].

Already observation-based estimates show that rapid warming of the ocean and rising sea level of the Earth's oceans over the past few decades contributes to increases in hurricane strength (e.g. three category 5 Atlantic hurricanes in 2017), hurricane rainfall (e.g. hurricanes Harvey in 2017 and Florence in 2018), storm surges, floods, the destruction of coral reefs, and

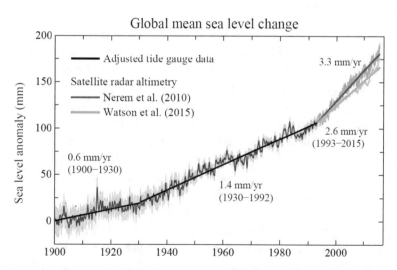

Fig. 5.7 Mean change of in global sea level from 1900 to 2015 based on tide gauge data (https://climate.nasa.gov/vital-signs/sea-level/) and satellite radar altimetry [35, 55]. Note that the rate of sea level change has been accelerating with time because of the increasing rates of ice sheet melt (Fig. 5.9b) [36]. For example, the Earth has lost 28 trillion tons of ice in the last 23 years [52]. Figure source is https://climate.nasa.gov/news/3012/nasa-led-study-reveals-the-causes-of-sea-level-rise-since-1900/

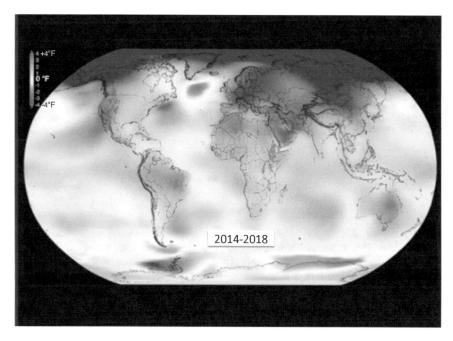

Fig. 5.8 Higher-than-normal air temperatures marked in red, orange, and yellow reveal a warming globe compared with the 1951–1980 average global temperature. The Polar Regions in the Arctic and Antarctic Peninsula seen in red in this temperature anomaly map, are on avereage 2.2 °C (4°F) warmer than 1951–1980 and warming at a rate 2 to 3 times faster than the global average [40]. In June 2020 temperatures in Siberia were up to 10 °C (18°F) higher than 1951–1980 (https://www.bbc.com/news/world-europe-53317861). The pattern and temperature increases shown in this figure are similar to those 3 to 5 million years ago when the CO_2 content was the same as now. Figure source is NASA Goddard Space Flight Center

global decline in ocean oxygen, ice sheets, glaciers and ice caps in the polar regions (Figs. 4.6, 5.2, 5.7 and 5.9) (https://en.wikipedia.org/wiki/Tropical_cyclones_and_climate_change). Higher than normal storm surges result because of the accelerating rate of sea level rise during this same period, the expansion of ocean water as it warms, and the greater melting of polar ice (Figs. 5.7 and 5.9) [20]. Also, ocean life is affected by warmer temperatures and by consequent lower oxygen levels in the ocean (Fig. 5.2) (https://en.wikipedia.org/wiki/Ocean_deoxygenation).

The consistent ocean warming and rising sea level, with little fluctuation, are the best evidence for global warming because the oceans absorb most of the Earth's increased heat (e.g. equivalent to the energy of three Hiroshima-sized atomic bombs every second) (Guardian newspaper calculation) (Figs. 5.2 and 5.7). The atmosphere has more temperature fluctuations, and short periods when the air temperature has not risen (Fig. 5.1). Climate

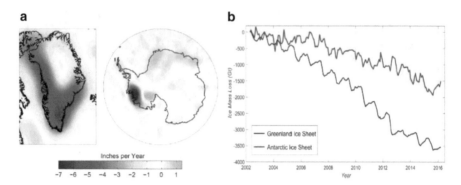

Fig. 5.9 **a** Accelerating polar ice sheet melt in Greenland (left) and Antarctica (right), which results in accelerating sea level rise (see Fig. 5.7). Water equivalent ice loss between 2003 and 2013 measured by the NASA GRACE satellites. **b** Accelerating cumulative mass volume loss from the Greenland and Antarctic ice sheets. Today, Greenland is contributing more to sea level rise than Antarctica, but this could change. Figure source is https://climate.nasa.gov/vital-signs/ice-sheets/ and 2018 update at https://svs.gsfc.nasa.gov/30880

change deniers cite these atmospheric temperature fluctuations as a lack of evidence for global warming. It is increasingly difficult to use air temperature fluctuation to deny global warming, however, because there is consistent ocean warming with little fluctuation since global records began in 1958 (Fig. 5.2). In addition, the past 5 years of ocean temperatures are the hottest on record (e.g. Fig. 4.6) [5].

The accelerating ocean temperatures affect sea level rise in two ways (Figs. 5.2 and 5.7). As the ocean warms, more sea ice melts and ocean volume expands. At the current global warming temperature rise of about 1 °C (2.12 °F) a sea level rise of 20–60 cm results over time (The Economist, August 17, 2019). If temperature rise is not kept below 2 °C (4.2°F) and more likely reaches greater than 3 °C (6.4 °F) without more radical reduction of greenhouse gases, thermal expansion alone will result in as much as a 30 cm rise of sea level [5]. This appears likely since present policies and the CO_2 reductions are not meeting goals of the Paris Climate Agreement and suggest that the average global temperature will rise slightly more than 3 °C (6.4 °F) (The Economist, August 17, 2019).

In view of the likely 3 °C (6.4 °F) rise of temperature, the IPCC [20] now estimates sea level rise will exceed 1 m (3 feet) by 2100 because of the growing contributions from Greenland and Antarctic ice melting (Fig. 5.9). An additional 1 m of sea level rise could be added by 2100 because of the present breaking away of large ice shelves (see Sect. 5.5). When these ice shelves flow away, unstable ice-edge cliffs higher than 100 m result, which cause additional cascading ice collapse into the sea, accelerated ice melting, and a possible 2 m sea level rise by 2100 or 3 m by 2200 [7]. It is important to remember that all these sea level rise numbers are the average rise and have the greatest effect in low lying coastal areas. There will be local variability, such as uplifted coasts, where sea level will not rise.

If the expected 1 or even possible 2 m rise of sea level occurs over the next century, many people wonder what the problem is when there is a long time to adjust to this sea level rise. One problem is that hurricanes already are more extreme and storm surges are higher because of ocean warming and sea level rise (Figs. 5.2 and 5.7). The combined effects have resulted in thousands of deaths in Asia and great economic loss. As well as the tragic loss of life, there are other human costs such as the relocation of 24,000 Vietnamese per year from the Mekong Delta [29] and 250,000 families to higher ground in Bangladesh at a cost of $576 million (The Economist, August 17, 2019). It cost $65 billion to repair 2012 hurricane Sandy storm surge damage in the New York City area. Venice has spent 5.5 billion euros ($6.6 billion) for flood barriers and as I write this text, it is undergoing historic floods, which will occur every day with a likely sea level rise of 50 cm. Another example of the present affects of sea level rise is that the London Thames Barrier for estuary floods closed just eight times between 1982 and 1990 and has closed 144 times since 2000. To prevent future damage related to sea level rise (Fig. 5.7), Miami Forida is investing nearly $500 million in pumps, New York City is spending $800 million to shield it from another Sandy, and Indonesia is building a $40 billon wall to protect Jakarta. These numbers are all small compared to the United Kingdom National Oceanographic center, which in 2017 estimated that the damage caused by sea level rise may reach as much as 14 trillion dollars a year by 2100 [29].

Compared to these examples of costs related to sea level rise, the scope of future global change problems related to higher sea level is enormous (Fig. 5.7). About 1.6 million kilometers of coastline along 140 countries face the sea (The Economist, August 17, 2019). Two thirds of the world's largest cities are located on these coastlines. A billion people live only ten meters above sea level and recent storm surges nearly 5 m high have killed thousands in Asia, Europe and USA. Many of the 680 million people around the world living in low-lying coastal areas will experience annual flooding events by 2050 that used to occur only once a century [20]. Also, flooding could happen at least once a year in many vulnerable cities by 2050. By 2100 the value of property on the worlds coastlines is estimated to between $20–200 trillion (The Economist, August 17, 2019). In USA alone by 2100, 2.5

million coastal properties worth \$1.1 trillion could be at risk of flooding every two weeks. In just New York City, 72,000 buildings worth \$129 billion sit in flood zones.

5.5   Diminishing Polar Ice, Mountain Glaciers, Permafrost

The following paragraphs show some of the changing patterns of sea ice, ice shelves, glaciers, permafrost and frozen methane (CH_4) (gas hydrate) that are related to global warming. All types of sea and glacial ice, snow cover permafrost, gas hydrate, ocean temperature and biota data show significant changes because of global warming, especially in the Arctic and Antarctic (Figs. 5.8, 5.9 and 5.10). I will focus mainly on the Arctic because I have conducted different types of environmental research there from 1966 to 2000 and have spent a dozen field seasons working in the Arctic studying both offshore on oceanographic cruises and onshore. The main source of data in the following paragraphs is from the 2018 Arctic Report Card, sponsored by NOAA (USA National Oceanographic and Atmospheric Agency) [40]. Some 85 researchers from a dozen countries contributed to this yearly report now in its 12th year.

Observations of the Arctic temperature, snow cover, sea ice and glaciers show significant effects of global warming (Figs. 5.8, 5.9 and 5.10). The Arctic snow cover has been below average for 11 of the past 12 years. The loss of snow cover is important because it reflects most of the sun's warming, whereas the increased darker ocean and land surface absorb most of the sun's heat and result in Arctic temperature amplification (Fig. 5.10). Similar to snow cover, historic data show the current observed rate of sea ice decline and warming temperatures are higher than at any other time in at least the past 1,500 years and likely much longer. Declining Arctic sea ice in 2017, with the maximum winter sea ice area measured each March, is the lowest ever observed (e.g. Fig. 4.10). Sea ice is also getting thinner each year, with thinner ice covering a wider area. Like the day-to-day weather, there are fluctuations year by year in sea ice cover. Consequently, the significant global warming proof is the long-term pattern of 13% decrease of sea ice per decade shown by decades of NASA (USA National Atmospheric and Space Agency) satellite photos.

Personal observations support the aforementioned scientific data. Jeremy Mathis, director of the Arctic Research Program for NOAA, told CNN in October, 2017 that he had witnessed an "extraordinary transition" in the

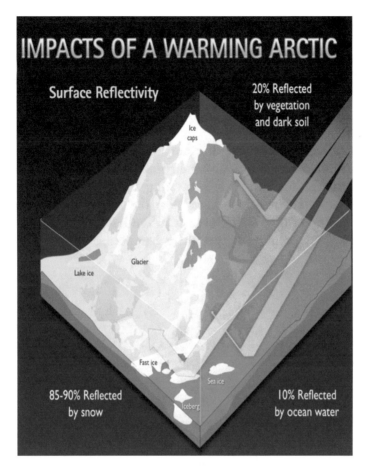

Fig. 5.10 Surface reflectivity of the suns incoming radiation shows that 85 to 90% of the suns heat is reflected by snow cover on land and ice, whereas 80% is absorbed by land surface and 90% is absorbed by ocean water. The global warming loss of snow cover on land and over sea ice, particularly in the Arctic, contributes to the amplified global warming in Polar Regions (see Fig. 5.8). Figure source is Corell [61]

Arctic environment since 2003, during which time he has made 14 trips to the region. "When I started going to the Arctic in 2003, it was very different environment than it is today," he said. "Back in 2003, we were breaking ice everywhere we went, from pretty much starting in the Bering Strait moving all the way up into the study areas north of Alaska and into the central Arctic basin. There was ice that was very thick and it was very extensive. "This year in 2017, during a 25-day cruise in the Arctic, we didn't see a single piece of ice (Fig. 4.10). We were sailing around on a coast guard icebreaker in blue water that could have been anywhere in the world. And it certainly didn't look like the Arctic." The Arctic environment has changed from one that was dominated by sea ice "to one that is now largely ice-free during the summer months," Mathis said. "And that transition in ice cover has had real impact on everything from fish to polar bears because everything in the ecosystem is really built around that ice coverage." For example, see my discussion about how the significant loss of sea ice cover has affected the Pacific walrus population in the Bering and Chukchi Seas (see Chap. 4 Sect. 4.5) (Figs. 4.6, 4.10 and 4.11).

Antarctica has a global warming story similar to the pattern of the loss of sea ice from rising temperatures in the Arctic. In Antarctica there is not only sea ice, but also glaciers covering the continent and ice shelves extending seaward from the glaciers, like the Larsen Ice Shelf along the Antarctic Peninsula (Fig. 5.11). Similar to the loss of sea ice in the Arctic, McClintock [28] points out that there has been a rapid retreat of and

Fig. 5.11 Map of Antarctica showing Antarctic Peninsula, Larsen Ice Shelf, the East Antarctic Ice Sheet (EAIS), the West Antarctic Ice Sheet (WAIS) and Wilkes Land. Map source is https://www.grida.no/resources/5339

decreased duration of sea ice, which annually doubles the size of Antarctica each year. He also mentions that there has been a 40% reduction of sea ice along the Antarctic Peninsula during the past 30 years, where the effects of global warming have been most severe (Figs. 5.8 and 5.9). The relatively greater temperature increases in Antarctica, compared to temperate regions, have resulted in the loss of many of the ice shelves and blocks of ice breaking off from the shelves such as one the size of Delaware that recently broke off from the Larsen Ice Shelf (Fig. 5.11) (National Geographic 2017). This huge iceberg, similar in size to South Georgia Island, is now on a path to collide with it and cause significant ecological damage https://www.theguardian.com/environment/2020/nov/04/giant-antarctic-iceberg-on-collision-course-with-british-territory-of-south-georgia).

The East (EAIS) and West Antarctic Ice Sheet (WAIS) continental glaciers on land contain 89% of the world's glacial ice and also have been affected by climate warming (Fig. 5.11) [57]. The East Antarctic Ice Sheet (EAIS) contains the greatest volume of ice (about 75%) and has been considered the most stable part of the Antarctic glaciers. However, there is new geological evidence for ice margin retreat or thinning in the vicinity of the Wilkes Sub-glacial Basin of East Antarctica during warm late Pleistocene interglacial intervals. At that time Antarctic air temperatures have been at least two degrees Celsius warmer than pre-industrial temperatures for 2,500 years or more. Consequently there appears to be a close link between extended Antarctic warmth and ice loss from the Wilkes Sub-glacial Basin. This ice loss suggests that in the future there may be a contribution to sea level rise from a reduced EAIS if an extended period of global warming continues (Fig. 5.9 B).

Recent studies show that at present, there is increased mass loss of ice in Wilkes Land part of the EAIS (Fig. 5.11) [44, 50]. As mentioned previously, increased ice mass loss has been found by studies predominantly in west Antarctica and the Antarctic Peninsula. The newly discovered increased mass loss in Wilkes Land suggests that the ocean warming already may be influencing ice dynamics in the marine-based sector of the EAIS. Ongoing warming in the Southern Ocean may adversely impact this region plus other parts of the Antarctic ice sheet, and this process may cause ice destabilization. The bottom line is that the change in the mass balance of the total Antarctic ice sheet from 2008 to 2015 decreased by 54% between 2008 and 2015 and its mass loss tripled between 2007 and 2016 (Fig. 5.9b) [20]. These observations show that mass loss from the Antarctic ice sheet is accelerating and has consequences for sea level rise.

The most recent study now shows that Antarctic ice is melting six times faster than it did in the 1980s (Fig. 5.9b) [[44]]. Since 2009, almost 278 billion tons of ice has melted away from Antarctica per year, whereas in the 1980s, it was losing 44 billion tons a year. Over all global warming has melted more than 3 trillion tons of ice in Antarctica since 1992. This has caused global sea levels to rise a half an inch since 1979, which does not sound like much, but it's a preview of things to come. As the Antarctic ice sheet continues to melt away, a multi-meter sea level rise from Antarctica is expected in the coming centuries, and a 3 m (10 feet) rise is possible this century. In comparison to the Antarctic ice loss, the Earth's total ice loss is 28 trillion tons for the 23 years from 1994 to 2017 (https://www.sciencealert.com/earth-lost-a-staggering-28-trillin-ton nes-of-ice-in-23-years-scientist-find).

Along with glaciologists worldwide, I also have personally witnessed the same pattern of significant loss of ice from mountain glaciers. Since my visits to Glacier National Park in USA and Pyrenees in Spain during the 1960s, many of the glaciers have disappeared. Similar to my personal observations, nearly all 1500 mountain glaciers have been receding worldwide in the twenty-first century (New York Times, Vanishing Ice Series, February 15, 2019). In contrast, when working in Argentina in 2017, I learned that some of the spectacular glaciers in Patagonia were advancing. This made me think, the global warming deniers would love this. However, I learned that the recent advance of a few Argentine glaciers has been because of the rapid ice thawing from global warming. These advancing glaciers have lost so much ice volume that they have become thinner and are now flowing faster and moving forward.

Recent studies indicate that glaciers of south and central Asia have 50% less ice than previous estimates and that 60% of the mountain glaciers will be gone by 2060 [21]. The loss of ice has been accelerating and ice volume has decreased 3% each year in 2017 and 2018 (New York Times, Vanishing Ice Series, February 15, 2019). This ice loss has significant implications for loss of human water supplies and less generation of hydroelectric power. For example, Switzerland generates 60% of its electricity from hydroelectric power that mainly results from glacier melt-water flow. Unfortunately, all mountain glaciers are estimated to disappear from Switzerland by 2090. To compensate, Switzerland has already begun funding construction of new hydropower plants to utilize water flow from large lakes that are being formed by the accelerated melting of the mountain glaciers.

In all the mountain glaciers, sea ice, ice shelves and continental ice sheets, the significant global warming pattern, especially in the Arctic, is the loss of ice volume that is accelerating (Figs. 4.10, 5.8 and 5.9). This is also true of the glaciers covering Greenland, which make up 10% of the Earth's glaciers. The Greenland ice sheet continued to lose mass or ice volume in 2017, as it has since 2002 (Fig. 5.9) (https://en.wikipedia.org/wiki/Greenland_ice_sheet). A study, based on data between 2003 and 2008, reports an average melting trend of 195 cubic kilometers (47 cu miles) per year and the ice melt rate became four times higher from 2003 to 2013. The concern over the melting of the Greenland ice sheet is that it is the major contributor to the continuing rise of sea level. There are many studies of Greenland and other ice melting that indicate a minimum of about a meter (about 3 feet) of sea level rise by the end of the twenty-first century (Fig. 5.7) (e.g. Watson et al. [55]). The worst case studies suggest an extreme 7 m rise (about 22 feet) over centuries to millennia if global warming continues at the present rate and the Greenland ice sheet melts (https://en.wikipedia.org/wiki/Sea_level_rise). This would result in flooding of most of the coastal cities of the world.

Global warming not only melts ice, but it also results in melting of frozen methane (CH_4) below the seafloor and frozen permafrost ground on land. This is especially true in the Arctic where the temperatures in the air and water have increased the most and permafrost is melting 70 years sooner than predicted (Figs. 4.6, 4.9, 5.8 and 5.9) [43]. The reason this is important is that when the frozen CH_4 melts, the gas is released into the atmosphere and CH_4 is responsible for 25% of global warming [32]. Permafrost and frozen CH_4 degrade on warming and thus rapidly increasing releases of CH_4 from these sources have been measured as a result of Arctic temperature amplification (Figs. 5.6 and 5.8). For example in 2013, land-based permafrost in the Siberian arctic was estimated to release 17 million tons of CH_4 per year, which is a significant increase from the 3.8 million tons estimated in 2006 and just 0.5 million tons before then (Fig. 3.2) (https://en.wikipedia.org/wiki/Arctic_methane_emissions). This compares to around 550–650 million tons released into the atmosphere annually from all sources [32].

The reason for concern is that this rapid increase in CH_4 release in Siberia is just part of the vast Arctic area where the frozen CH_4 is found (Fig. 3.2). In the 1970s, some of my USGS colleagues mapped extensive areas of frozen CH_4 under the seafloor of Bering and Arctic seas. It is estimated that at least 1,400 gigatons of carbon is presently locked up as frozen CH_4 [51]. Because 5–10% of that area is subject to large episodic eruptions of methane, that have been observed [34], they conclude that release of up to 50 Gt of the predicted amount of the frozen CH_4 storage is possible for abrupt release at any time. That would increase the methane content of our Earth's atmosphere by a factor of twelve.

The warming of the ocean, melting sea ice and rise of sea level not only cause melting of frozen methane, but these combined effects increase erosion of coastal permafrost that can result in an additional release of CO_2 into the atmosphere. The melting of permafrost is accelerating along Arctic coast and permafrost is eroding into the sea. As sea ice-free conditions expand in the Arctic, cliffs and shorelines are exposed to storms and wave action for longer periods, which accelerates erosion. The Arctic coastlines with permafrost make up more than a third of Earth's coasts. These coasts are eroding at an average rate of roughly half a meter per year, although in some spots the rate tops 20 m per year (EOS, November 15, 2019). Climate models assume that the organic carbon in eroded permafrost is consumed in primary production or buried offshore. However a new study suggests that between about 1% and 13% of the initial total organic carbon is vented back into the atmosphere as CO_2 or other greenhouse gases [53].

This newly recognized release of CO_2 is a concern for global warming because the permafrost region covers roughly a quarter of the land in the Northern Hemisphere (EOS, November 15, 2019). This permafrost stores more than double the amount of carbon in the atmosphere today, with much of it still locked away, frozen. Scientists have known for some time that permafrost could become a major source of greenhouse gases as the soil thaws and once-dormant microbes wake up and break down organic matter. A new study confirms other aspects of Arctic temperature amplification, because accelerated global warming in the Arctic, reduced snow and ice cover, melting permafrost, and coastal erosion are adding CO_2 to the atmosphere (Fig. 5.8) [53].

5.6 Past 20 years of Record-Breaking Temperatures

There has been a pattern of 16 of the global record breaking average near-surface air temperatures occurring since 2001 (Fig. 5.1). Nearly every year has set a new record, except for years with exceptionally strong El Niño's when temperatures are even higher [15]. Overall since 1880 when good instrumental records began, the global average temperature has risen 0.85 degrees Celsius or °C (1.8 °F) with the majority of that increase since 1980 (Fig. 5.1). The hope of the Paris Agreement is that average global temperature rise will be kept below the 2 °C (~3.6°F). The Intergovernmental Panel on Climate Change (IPCC) suggests that with this limit, perhaps the worst effects of global warming can be avoided. The average global temperature in 2017 was

+ 1.17 °C (~2.1°F) relative to the average temperature for 1880–1920, which Hansen et al. [15] take as an appropriate estimate of pre-industrial tempera- ture. The 2017 temperature and continuing average temperature warming in the air and ocean shows that we already are approaching critical temperature increases (Figs. 5.1 and 5.2).

Most troubling, are the greater record setting temperature patterns observed in Polar Regions (Fig. 5.8). For example, the Arctic Report Card [40] shows that there was a warmer Arctic average annual air temperature over land in 2017 and it was the second warmest year since 1900. The tempera- ture was 1.6 °C (2.9 °F) above the average for the period of 1981 to 2010. Satellite data showed trees taking over grassland and greener Arctic tundra with bigger plants and leafier shrubs. There also was an above-average ocean temperature, with sea surface temperatures in August 2017 in the Barents and Chukchi seas up to 4 °C (7.2 °F) warmer than average (Fig. 4.6). As a result there were Increased Arctic ocean plankton blooms, because retreating sea ice allowed sunlight to reach the ocean and encourage more marine plant growth across the Arctic.

All of the aforementioned data of the Arctic Report Card [40] show that the Arctic is warming more than twice as fast (+2 °C) (3.6 °F) than the rest of the Earth (+0.85 °C) (1.5°F), a phenomenon known as Arctic tempera- ture amplification (Fig. 5.8). Because of the loss of snow cover and ocean ice sheets, the Arctic is changing from a highly reflective surface to a very dark surface. This means that sunlight entering the atmosphere, which would normally be reflected back into space by Arctic sea ice or snow and glaciers, is instead being absorbed by dark surfaces like the ocean, land and rock (Fig. 5.10). Because of the greater absorption of sunlight heat, the global warming of the Arctic is amplified by this feedback process.

Although this Arctic warming has positive effects for plant growth, this is outweighed by the negative effect that melting permafrost has for global warming. Permafrost covers 16.7 million square kilometers (6.45 million square miles) or one fourth of the northern hemisphere land mass [48]. From my experience in the Arctic, I can confirm that as permafrost melts, it becomes a sponge of organic debris. Because of this organic content, as the permafrost melts, it releases both carbon dioxide and methane greenhouse gases. Recent measurements indicate that as a result of permafrost melting, 130–160 billion tons of carbon, mainly CO_2, will be released during this century [48].

The same temperature amplification patterns observed in the Arctic are also observed in Antarctica. McClintock [28] notes that the Ukrainian

Research Station in the Antarctic Peninsula has measured winter temperatures for 60 years and found temperature has increased 6.1 °C (11 °F). Like the Arctic, these Antarctic temperature increases are much greater than the increases of the global average temperature (Fig. 5.8). Long-term climate change isn't just captured by thermometers, such as by the Ukrainians in Antarctica. Grapes harvest dates going back to 1354 in the Burgandy country of France also have been used to reconstruct a 664-year record of temperature traced by fruit ripening [24]. Grape harvest dates are a proxy for summer warmth. Grapes have been, on average, picked on 28 September or later before 1988. However, from 1988 onward, grapes have been harvested roughly 13 days earlier. Hot and dry years in the past were unusual, while they have become the norm since the transition to rapid warming in 1988.

In the following paragraphs on weather extremes intensified by global warming, I will show how all the humans on the planet are going to feel the impacts of global warming, particularly in the Arctic and Antarctica (Fig. 5.8).

5.7 Worse Extreme Weather Events

A pattern of increasingly severe hurricanes, floods, tornados, droughts, wildfires, and cold weather events is occurring worldwide, some of which can be linked to global warming (Figs. 5.12, 5.13 and 5.14). However, there is scientific debate about the link to the changes in the jet stream that are causing some of the extreme weather events (e.g. Masters [27]). Some scientists say the changes are a direct result of global warming and others say these changes of the past 15 years are not long enough to prove this link to climate warming (e.g. Francis and Vavrus [10]; Seviour [49]). For example, the extreme rainfall of hurricanes Harvey in 2017 and Florence in 2018 were caused by jet stream changes that slowed the movement of the hurricanes so that they dumped much more rain than usual (e.g. Fig. 5.12). The link to global warming now may be getting stronger because new improved climate models predict that these jet stream changes blocking hurricane movement are expected because of climate warming (https://archive.thinkprogress.org/hurricanes-like-harvey-are-the-new-normal-dd508006902e/).

There have always been severe weather events and you cannot take one single weather event like a hurricane or a period of drought or less ice one year in the Arctic as proof of climate change. However there is evidence for a global warming cause when a worldwide pattern of increasing and more severe events is observed and it can be linked to specific trends of global

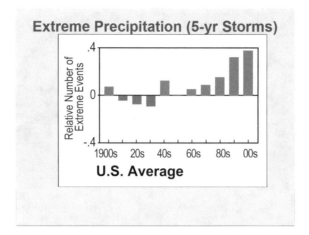

Fig. 5.12 Increase in frequency of two-day heavy rain events defined as one that occurs on average only every five years. Percent changes are compared to the period 1901 to 1960 and do not include Alaska or Hawaii. Note that there has been a large increase in the number and size of extreme precipitation events (e.g. hurricanes Harvey 2017 and Florence 2018). Figure source is [22]

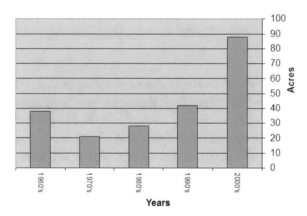

Fig. 5.13 Average size in acres of United States wildfires is shown by decade from 1960 to 2009. Both the number of fires and size has increased dramatically since 2009. For example, California had the three largest (burning over one million acres) and deadliest fires in history in 2018 and in 2020 only part way through the fire season, there has been a 36% increase in the number of fires compared to the previous year (fire.ca.gov). Figure source is https://wildfiretoday.com

warming and the extremely rapid increase of CO_2, CH_4 and temperature compared to the past several hundred thousand years (Figs. 5.3 and 5.6). For example, climate change does not cause severe storms like hurricanes or monsoons. However, global warming results in consistently warming oceans

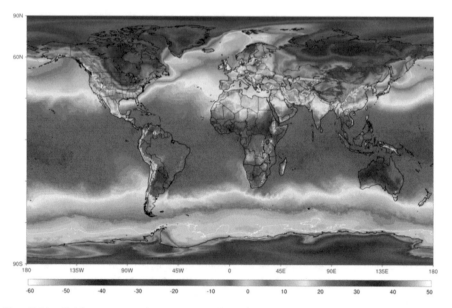

Fig. 5.14 Cold temperature extremes for the northern hemisphere on January 30,2019, that were caused by the southward expansion of the polar vortex (shown in purple color). At the same time contrasting high temperature extremes for low latitudes (shown in red color) caused extreme droughts. Figure source is https://cli matechange.umaine.edu

and rising sea levels, which intensify extreme storms (4.2, 4.7). Consequently storms have greater wind speeds (e.g. three category 5 Atlantic hurricanes in 2017), rainfall (e.g. greater than 5 feet of rain in Houston from hurricane Harvey, 2017) and storm surge heights several meters higher than normal because of stronger winds, larger waves and higher sea level (Figs. 5.7 and 5.12). Because of the warmer ocean water and more water vapor in hurricanes, even lower category hurricanes like Florence dump abnormally large amounts of rain, e.g. several feet and category 5 hurricanes like Harvey cause even greater floods [58].

These patterns of more intense storms are exemplified by hurricanes Andrew, Katrina, and Sandy, the catastrophic trio of Harvey, Irma and Maria hurricanes in 2017, and Florence in 2018. After Katrina, 39% of the USA population believed climate change fueled more intense storms; however after the 2017 hurricanes, 55% believed climate change was the cause [26]. Hopefully, science is starting to win the battle over the propaganda of climate-change deniers. This is important because of the catastrophic storm losses of thousands of lives and billions of dollars of damage just caused by USA hurricanes in 2017 and 2018. For example, hurricane Florence in 2018 combined with 2017's triumvirate of devastating hurricanes Harvey, Irma, and Maria resulted in a combined death toll of 3,100 and damages estimated to be $275 billion [58]. Wing et al. also point out how FEMA studies have greatly under-estimated USA flood prone areas, which was evident in hurricane Florence. It is important to note that these studies were conducted in Britain, which is another example of how the lack of science funding is hurting USA federal science agencies. These catastrophic losses will only get worse in the future if the USA government does not study and address the fundamental causes and risks of global warming.

Unfortunately, the government interest to combat global warming is more difficult in USA than most other countries. Many in USA, like President Trump, use the extreme cold weather around New Year's Eve in 2017 or January 2019 to support their denial of global warming, when actually extreme cold weather may be intensified by global warming (Fig. 5.14). Previously I mentioned how the loss of snow and ice cover is resulting in an increasingly dark and warm earth's surface in the Arctic. Because of the darker, warmer land and water surface and greater heat absorbed, the polar vortex of cold Arctic air may not be confined as well within the Arctic Circle as it previously was (Fig. 5.10) (e.g. Francis and Vavrus [45]) (https://en.wikipedia.org/wiki/Polar_vortex). As a result, more and larger breakouts of Arctic air masses are taking place further into the central, eastern and southern USA. The 2017 and 2019 severe cold weather may be an example of this.

I certainly observed severe cold weather living for 25 years in Minnesota from 1937 to 1962. However, I never recall snow as thick as that covering my brother's one-meter high mailbox or the historic low temperatures of minus 48 °C (-55 °F) in northern Minnesota a few years ago. This anecdotal evidence agrees with predictions that extreme cold weather from Arctic vortex breakout episodes will become more frequent with global warming, such as the coldest temperatures on record in central USA in 2019. Also, predicted and supporting the link to global warming are the observations that during the 2017 extreme cold weather pattern in central and eastern North America, much of the rest of the world in mid-latitudes experienced unusually warm weather, even in Alaska (Fig. 5.14).

In contrast to the severe cold weather events, I have experienced extreme hot weather, El Niño's, droughts and associated massive wildfires in the past few years in Oregon, California and Spain and these have been equally bad globallly, such as in the summer of 2019–2020 in Australia (see more details in Chap. 6 Sect. 6.4) (Fig. 5.14). These are another set of extreme weather patterns that are intensified by the higher temperatures of global warming, which make droughts worse, wild fires larger and the wildfire season about

80 days longer in USA, because of lower soil moisture (Fig. 5.13) (The Economist July 17, 2019).

In USA alone, a new record of 10 million acres (4,046,856 hectacres) of forest burned in 2015, and California experienced its deadlist and largest wildfires in 2018 and 2020, because of warmer temperatures and lower soil moisture. Six of ten of the most destructive wildfires in California history have occurred during the last decade. This wildfire cost has been $3.8 billion for fire fighting, more than the total for the past 30 years (The Economist November 17, 2018). Wildfires consumed more than 3.5 million hectares in the United States in 2018, destroyed more than 18,000 residences, and federal fire suppression costs topped 3 billion dollars (Levi et al. 2019). A major contribution to the increased intensity of wildfires is that 100 million trees have died during the past decade of droughts in California and these help fuel the wildfires [29].

Unlike the immediate devastation of drought and wildfires, there are also long-term climate changes that affect millions of people and have huge economic costs. For example a recent USGS study shows that global warming and drought have been shrinking the Colorado River water flow [33]. This flow provides water to more than 40 million people from Denver to Los Angeles and supports more than $1 trillion per year of economic activity. The study found that more than half of the decline in the river's flow is connected to increasing temperatures, and as warming continues, the risk of severe water shortages for millions is expected to grow. Without any cuts to greenhouse gas emissions (e.g. CO_2 and CH_4), the river's discharge could shrink by between 19 and 31% by the middle of the twenty-first century.

In contrast to worse droughts, the extreme El Niño of 2015–2016 caused torrential rains and floods in places like California and England, as well as droughts and wildfires in Indonesia that blanketed Southeast Asia in smoke (Fig. 5.12). The extreme El Niños are expected to double if global warming continues unabated [9]. The drought-related events have been most severe in the world's dryland areas. Drylands are areas where evaporation is greater than precipitation. Drylands cover 41% of the earths land surface and contain 38% of the world's population [17]. The increasingly severe temperatures and droughts in these areas have resulted in desertification, which is the increase in the earth's desert areas (Fig. 5.14). A 2017 study by East Anglia University estimated that if average global temperatures rise 2 °C (4.2 °F) by 2050, severe drought and desertification will result over 25% of the earth [29].

One of the worst examples of desertification is occurring in the Sahara region of northern Africa. During the past few decades the Sahara desert has expanded 250 km to the south, 70% of the of the Sahara arid area has deteriorated and water resources have disappeared (Fig. 5.14) [17]. As a result of droughts, loss of water resources and related overgrazing, farmland has been lost and six million refugees have fled the Sahara area. This has contributed to the immigration problem from Africa to Europe.

The worst drought cost is famine and the loss of human life. Droughts are a natural hazard that during the last century has caused one of the greatest losses of human life in the millions. For example an estimated 3 million lives were lost in China in 1923, nearly 2 million in India in 1943 and nearly 500,000 in Ethiopia and Sudan in 1983 (https://www.statista.com). Better famine relief programs and increasing migration from drought areas are lowering deaths from famine. In the future, however, it can be expected, that there will be more deaths, migrations, climate change refugees, and conflicts from increasingly severe droughts intensified by global warming. Overall, although extreme weather events can cause more rainfall and greater plant growth, the extreme droughts outweigh the positive effects of increased rainfall. The bottom line is that both extreme weather droughts and floods, intensified by global warming, result in increased loss of human life and economic destruction.

An example of the combined effects from the loss of water and food resources took place in 2010 when an atmospheric jet stream change, that appears to be related to global warming, resulted in severe drought in Russia and the Middle East [27]. In addition, low atmospheric pressure was pushed into Pakistan and Bangladesh where catastrophic floods resulted. The heat wave in Russia killed 55,000 people and destroyed much of the wheat crop. The loss of wheat increased global grain prices and contributed to the Arab Spring conflicts of 2011. Even worse, in Syria the rapid reduction in ground water in 2010 and previous years, decreased food production even more. Global warming consequently has contributed in part to the civil war in Syria that has continued until the present and resulted in the immigration conflicts from the millions of refugees in the Middle East and Europe.

In sum, mainly because of the worse extreme weather events related to global warming, there has been a significant increase in civil disorder and natural disasters during the past 20 years. Between 2000 and 2019, there were 7,348 major natural disasters, compared to the 4,212 disasters recorded from 1980–1999 (United Nations, 2020, https://www.undrr.org/publication/human-cost-disasters-overview-last-20-years-2000-2019). These disasters, including earthquakes, tsunamis and hurricanes, claimed 1.23 million lives, affected 4.2 billion people and resulted in $2.97 trillion in global economic losses during the new millenium. The vast majority of those disasters were climate-related, with researchers reporting more flooding, storms, droughts, heatwaves, hurricanes and wildfires.

5.8   Increasing Extinction Rates

In your home area have you noticed a loss of songbirds, which are one of nature's most enjoyable aspects? I have seen my favorite beautiful-colored warbler disappear in the past few years from my Spanish neighborhood. Recently I returned to my Silicon Valley neighborhood after 20 years and found it dominated by crows with hardly a songbird. This is not just anecdotal because previous and new research documents a significant loss of songbirds and other bird species. A new study finds that the United States and Canada have lost 2.9 billion birds, or 29% of 1970 abundance during the past 50 years [45]. Also a new report from the National Audubon Society estimates, that nearly two-thirds of the 604 North American bird species they studied, will go extinct if global warming hits 3 °C (5.4°F). Recent studies in Great Britain show that 25% of songbirds face extinction or steep declines in population [16]. Fortunately some British birds are surviving global warming by moving northward, but this is not an option for polar birds. However it is important to realize that climate change is not the only cause involved, because there are also other combined effects of global change such as loss of habitat, pesticide use, domestic cats and glass buildings.

Most of the same combined environmental problems of global change are affecting the insect population, which in turn affects bird populations. For example some of the insect eating birds in Great Britain have already declined by 77–93% as pesticide use has doubled since 1990 [12]. As pesticide use, habitat destruction and global warming increase, there also has been a 77% decline in butterflies in Great Britain. A summary of 73 studies, mainly from Europe and North America suggests that the rate of local extinction of insect species is eight times faster than that of vertebrates [46]. This summary estimated that, on average, insects are declining by 2.5% each year, with 41% of insect species threatened with extinction. Fortunately, it is not too late because few insects have gone extinct so far, and populations can rapidly recover [12].

Some people may say why does the loss of insects matter since many are just pests anyway. However, three quarters of our crops depend on insect pollinators and crops like strawberries will begin to fail [12]. Without insects 7.5 billion people cannot be fed. The value of insects for crop production is up to nearly half a trillion dollars annually. The biggest challenge to prevent the loss of insects is to make farming more wildlife-friendly. Pesticide reduction targets would help enormously. In Sichuan, China, farmers pollinate apple trees by hand because of the heavy use of pesticides.

From the loss of songbirds and insects, it is evident that climate change and human activities are already significantly altering the natural world. The increasing extinction rates of animal and plant species result from rapid global warming combined with other human-caused global changes. A study by Wiens [56] did not expect to see much change in species extinctions, because the world has just warmed 0.85 °C (1.5 °F) between 1880 and 2012. He investigated 976 plant and animal species worldwide in Asia, Europe, Madagascar, Oceania and North and South America as well as from freshwater, terrestrial and marine environments. His study revealed that 47% of plant and animal species have experienced local extinctions due to climate change. This does not mean that species have become extinct because the effects are local. Amphibian species that once frequented particular ponds and streams have slipped away, meadow wildflowers have migrated, and once-familiar butterflies and bees have flown favorite nesting places, all in response to global warming. The effect was more pronounced for tropical species, and most pronounced of all in animals: 545 out of the 716 species. For example there has been a 60% decrease of vertebrate species since 1960 (Fig. 5.4f).

A 2017 study on vertebrate species said the current extinction period, known as the Holocene extinction event, might be the greatest event in the Earth's history and the first due to human actions [4]. For example, in the past few decades the lion population in Africa has declined from 400,000 to 25,000 and there have similar declines in leopards, cheetahs and elephants [37]. Unlike previous events, however, extinctions are happening over the course of decades rather than centuries. Recent studies suggest that a quarter of the world's species may go extinct by 2050 (https://www.dw.com/en/worldwatch-publishes-state-of-world-report/a-15874848). An excellent summary of the extinctions and environmental changes in just a lifetime are summarized in a Netflix movie *David Attenborough: A Life on Our Planet* (https://www.neflix.ctom/es-en/title/80216393).

The most comprehensive recent study is not quite as gloomy as the Worldwatch report above [18]. The IPBS investigation estimates that one million of the eight million species will become extinct in the future and emphasizes that multiple global change factors cause the extinction of species. This IPBES study, which is written by 145 experts from 50 countries, indicates that more than 40% of amphibians, 33% of coral reefs and over a third of all marine mammals are threatened with extinction. Shrinking habitat, exploitation of natural resources, global warming and pollution are the main drivers of species loss. For example, 75% of land and 66% of marine environments have been altered since pre-industrial times, a third of marine stocks are fished at unstainable levels, plastic pollution has increased ten times since 1980, alien invasive species per country have risen by 70% since 1970, and global warming impacts 50% of mammals and 25% birds.

Biologists have warned for years that global warming could trigger not just local but global extinctions of animals and plants. This reduced diversity of species will occur especially in those species already threatened by habitat destruction, pollution, invasion of exotic species and overhunting. There are two important ways that global warming can affect species. First, most species are adapted to specific temperature regimes. When temperatures warm rapidly, species may not be able to survive. For example there have been massive die offs of alewife fish in the Great Lakes because of warming waters [8]. Another way species have adapted to warming is to migrate to other regions with the proper temperature regime, but sometimes this new habitat is less suitable and reduces the species population.

The second effect of rapid global warming is the reduction of habitat and food supply for a species, particularly in Polar Regions (Figs. 5.3 and 5.8). The great reduction of sea ice in Polar Regions significantly reduces habitat and food for many species (Fig. 4.10). This is particularly important for mammals such as walrus and polar bears, fish, and birds such as penguins (Fig. 4.11). McKlintock (2012) shows that there has been the loss of 12,000 of the 15,000 Adeline breeding penguin pairs at one location on the Antarctic Peninsula. This loss is caused because these penguins main food is Antarctic krill. The larval stages of the krill live under the sea ice and feed on the abundant algae on the bottom of the ice. As the sea ice disappears, so does the krill food source for the penguins as well as many other species of birds, fish and mammals in the Antarctic. A similar fate may take place for the world's most productive fishery in the southern Bering Sea. Again in this location, my colleagues at the University of Alaska have shown that productivity of the algae on the underside of the ice helps sustain this fishery. The loss of the annual Bering-sea ice, that I have previously described, will most likely reduce the productivity of this important fishery (Fig. 4.10).

Why are these future impacts of human-caused global changes on species extinction, biodiversity and ecosystem functioning important for us? First of all, we humans are a biological species and are also subject to extinction caused by multiple global changes including global warming. Also biodiversity and ecosystem functioning provide substantial benefits to people. USA Senator Udall notes, that among other benefits, at least 40% of the world's economy is based on biological resources and that diversity of life provides humanity with shelter, medicine, economic development, and food such as from the Bering Sea fisheries,.

These multiple patterns of worldwide change verify that global warming is already taking place and affecting all environments as well as plant and animal species on earth, including humans (Figs. 5.1, 5.2, 5.3, 5.4, 5.5, 5.6, 5.7, 5.8, 5.9, 5.10, 5.11, 5.12, 5.13 and 5.14). However, these global warming changes need to be put in perspective so that people do not throw up their hands in despair and do nothing because they think it is impossible to alleviate global warming. Some climate change alarmists have made statements that human life will end in a decade or two, but this is neither true nor beneficial [13]. Most of the world is taking steps to lower CO_2 emissions, even if it is not fast enough. The world does have a record of many cooperative agreements to prevent environmental damage (see Chap. 7, Sect. 7.4). It is important to remember that many other human-caused global changes are occurring that combine with global warming to cause the environmental problems and that warming alone cannot be blamed. In addition, natural earthquake, volcanic and weather disasters will always affect human's, however global warming will make extreme weather worse.

We need to remember that there might be some beneficial effects to global warming. A study of NASA points out that in the southern Sahara, 300,000 square kilometers of the Sahel are turning from desert into grassland. From a quarter to half of Earth's vegetated land has shown greening from 1982–2015 largely because of rising levels of atmospheric CO_2, which increase photosynthesis and spur plant growth [60]. This study [60] showed that CO_2 fertilization explains 70% of the greening effect. However, the beneficial impacts of CO_2 on plants also is limited, said co-author Dr. Philippe Ciais. He states "Studies have shown that plants acclimatize, or adjust, to rising CO_2 concentration and the fertilization effect diminishes over time." His statement is substantiated by careful studies at Duke Unversity. Experments with nearly 600 ppm CO_2, found pine forest grew only slightly better and increased growth may or may not take place depending on available nutrients in the soil [23].

To fully evaluate the positive affects of rising CO_2 on plant growth, these should be balanced against the negative affects of vegetation loss because of the increased areas of droughts from global warming. Also counteracting any possible increased plant growth is the affect of soot and smog in the air from fossil fuel and forest fire burning, which cause reflection/blockage of sunlight and reduce plant growth https://www.wef orum.org/agenda/2014/11/air-pollution-reduces-crop-yields/. The net affect of increased burning of fossil fuels and increased CO_2 is not beneficial overall for plants. So do not be fooled by some climate change skeptics who say burning fossil fuels and more CO_2 in the atmosphere has a net beneficial affect for CO_2 plant growth https://www.scientificamerican.com/article/climate-skeptics-want-more-co2/. In summary, neither uptake of CO_2 by plant growth or dissolving in the ocean water will solve the problem of human induced CO_2 [23].

5.9 Natural Versus Human-Caused Climate Change

Throughout Parts I and II and in this chapter, I have presented a number of my personal and other scientist's examples to show that the climate is warming (e.g. Figs. 4.6, 4.10, 5.1, 5.2, 5.8, 5.14). The most recent polls in the USA indicate that about 70% of the population believes the climate is warming and about 60% of these believe the post-industrial global warming is related to human activity [26]. However, about a third of the USA population believe global warming is caused by natural changes in the environment and about 40% deny or remain skeptical that humans are the cause. These data indicate that a review of human and natural causes for climate warming is important to confirm whether humans are the cause of warming. If so, then we humans, particularly political leaders, must take action to slow warming and reduce loss of life and economic costs.

So if the climate changed before humans, how can we be sure we are responsible for the dramatic warming that's happening today (<https://www.quantamagazine.org/how-earths-climate-changes-naturally-and-why-things-are-different-now-20200721/?>)? I will review the main natural and human caused controls on climate warming and then we can examine who contributes how much, i.e. natural earth temperature cycles or human activity. In this way we can determine if there is unnatural human-caused, post-industrial global warming. We will start with carbon dioxide (CO_2) since it is most cited as a contributor to human-caused global warming. Prior to the industrial revolution, the CO_2 content was about 285 parts per million (ppm) in the atmosphere (Fig. 5.3) (https://www.esrl.noaa.gov/gmd/news/7074.html). Since pre-industrial times of 1850, the amount of CO_2 has risen to over 400 ppm and the temperature has risen 1.28 °C (2.7 °F) (Figs. 5.1, 5.2 and 5.3) (http://berkeleyearth.org/archive/2019-temperatures/).

We know that over tens of millions of years the ocean temperature rise has been associated with increase of CO_2 (Fig. 5.5). We also know that humans have transferred 10 to 36 gigatons of carbon into the atmosphere per year from 1960 to the present because of burning fossil fuels (Fig. 5.4g). This rate of carbon release is about 10 times faster than anything scientists can find in the geological record for the past 300 million years (Lynas 2018, Alliance for Science at Cornell University). This natural carbon increase includes cataclysmic volcanic eruptions of CO_2 over millions of years that are linked with several of the mass extinctions of life at these times. An increase of nearly 100 ppm of CO_2 in the atmosphere has occurred in the last 70 years, whereas such a change without human interference has taken thousands to millions of years during geologic history (Figs. 5.2 and 5.3 ice age to warm periods, 4.5 e.g. Greenhouse to Icehouse world). In other words, the increase of CO_2 during the last century has occurred orders of magnitude faster than similar natural increases or decreases of CO_2.

The natural CO_2 content of the atmosphere has varied cyclically between ~ 180 and ~ 280 parts per million by volume over the past 800,000 years (Fig. 5.3) [54]. Although human output of 10–36 gigatons of CO_2 is tiny compared to the 750 gigatons moving through the natural carbon cycle each year, it adds up because the land and ocean cannot absorb all of the extra CO_2 (Fig. 5.4g) https://www.esrl.noaa.gov/gmd/news/7074.html). About 40% of the human produced CO_2 is absorbed. The rest remains in the atmosphere for up to 200 years after entering (https://archive.ipcc.ch/ipccreports/tar/wg1/016.htm). As a consequence, atmospheric CO_2 of 400 ppm is at its highest amount since Pliocene time 3 to 5 million years ago and global warming of the atmosphere and ocean has resulted (Figs. 5.2, 5.3 and 5.5 see red arrow).

Humans also are introducing methane (CH_4) into the atmosphere even faster than CO_2, and CH_4 is a 34 times more potent greenhouse gas than CO_2. (Fig. 5.6) (https://en.wikipedia.org/wiki/Methane_emissions) [32]. Like CO_2, the CH_4 has remained stable for the past 800,000 years, but now the CH_4 increase is a major concern because it is responsible for 25% of the present global warming temperature. Wetland activity is increasing rapidly because it is linked to global warming and this results in melting permafrost, thawing of frozen methane, which release increased CO_2 and CH_4 into the atmosphere.

From the above discussion, we can see that both CO_2, and CH_4 contents in the atmosphere have had little fluctuation between the ice age and warm periods of the past 800,000 years (Fig. 5.3). Without any human activity, this is what the climate would have been, and how it would have changed over the last and next 100–200 years. Now when we add in the human-caused element, we can see rates of change in CO_2, and CH_4 contents that are orders of magnitude faster than the natural fluctuations of the past hundreds of thousands to millions of years (Figs. 5.3, 5.5 and 5.6). Consequently, human-caused global warming, with all its ramifications, has resulted (e.g. see this chapter and Chap. 6) (e.g. Figs. 5.4, 5.7, 5.8, 5.12, 5.13 and 5.14). The warming, intensified extreme weather, wildfires and biological extinctions etc. will only get worse if humans do not reduce and eliminate their input of CO_2 and CH_4. Even if the input of all CO_2 and CH_4 is eliminated immediately the main effects of the present human input would last many years because it takes up to 200 years for CO_2, and 12 years for CH_4 to be eliminated naturally from the atmosphere (https://archive.ipcc.ch/ipccreports/tar/wg1/016.htm).

Natural activity from volcanoes has significantly increased CO_2 and warming of the climate in the past, however this not true now. For example, some geologists hypothesize that 250 million years ago an extensive flood of lava poured continually from the ground in Siberia for hundreds of thousands of years (Fig. 3.2) [30]. This large-scale, long-lasting eruption likely raised global temperatures enough to cause one of the worst extinction events in our planet's history. Large-scale volcanic eruptions during the rifting apart of the Atlantic Ocean appear to have raised CO_2 levels to 800 parts per million (ppm), or even higher and led to the greenhouse world of 45–55 million years ago (Fig. 5.5) [14].

Current volcanic activity and volcanic input of CO_2 doesn't occur at anywhere near the same massive scale as that of 10 s to 100 s of millions of years ago. Human activities now emit 60 times more carbon dioxide than the amount released by volcanoes each year [30]. Large violent eruptions, such as Mount Saint Helens in 1980 or Pinatubo in 1992, may match the rate of human CO_2 emissions for the few hours that they last, but they are too rare and fleeting to rival humanity's annual emissions at present. The massive volcanic eruptions, CO_2 input and natural warming of greenhouse worlds required huge scales of volcanic eruptions over hundreds of thousands to millions of years (Fig. 5.3 Greenhouse world) [14, 30]. There have been no volcanic eruptions on this scale for the past tens of millions of years. The only natural event that rivals the nearly instantaneous geologic increase of CO_2 at present and a rapid global temperature change is when a meteorite hit the earth 66 million years ago [1]. No such meteorite or large volcanic event has occurred during the last century to account for the unnatural CO_2 input or rapid temperature change we have experienced (e.g. Figs. 5.1, 5.2, 5.3 and 5.4g).

Several natural processes, that are not related to volcanic activity, are responsible for the warmer mid-Pliocene (3.3–3 million years ago) when CO_2

was about the same as the present, temperature was about 2–3 °C warmer and sea level about 20 m (65 feet) higher than now (Fig. 5.5) [31] (https://en.wik ipedia.org/wiki/Pliocene_climate). The warmer Pliocene climate in part has been attributed to the closing of the Ismus of Panama, greater salt in the Atlantic Ocean and more heat transfer to the north. Also, a nearly constant El Niño state resulted in a warmer equatorial Pacific Ocean, which caused more water vapor as a greenhouse gas in the atmosphere. During the Pliocene the earth climate system response shifted from a period of oscillation dominated by the 41,000-year period of earth's obliquity (axial tilt) to one of low frequency, high-amplitude oscillation dominated by the 100,000-year period of orbital eccentricity (deviation in Earth's orbit) characteristic of the Pleistocene glacial-interglacial cycles.

These natural changes of the closing of the Isthmus of Panama and climate cycles in the Pliocene took place over hundreds of thousands to several million years (Fig. 5.5). The extensive time that it took for the natural Pliocene increase in CO_2 and temperature is orders of magnitude longer than the similar changes of CO_2 and temperature have taken within the past century (Figs. 5.1, 5.2 and 5.3). The present global warming seems headed towards temperature increases of 2–3 °C (3.6–5.4 °F) in the next century that are similar to those of the Pliocene when sea levels were about 20 m (65 feet) higher (Figs. 5.5 and 5.15) (The Economist August 17, 2019). Fortunately, like the slow natural increase of polar ice, it also will take a similar long time for the present polar ice to melt and raise sea level. Consequently, even with global warming it will take millennia for sea level to reach 20 m (65 feet) higher, like the Pliocene.

I have mentioned that the tilt of the earth's axis and deviation in the earth's orbit affect natural variability of climate change, but what about the sun's energy that ultimately provides the entire natural warming for the earth's climate. A statistical method called fractional risk attribution or fingerprinting has shown that the rise in global temperature follows a pattern expected from human-caused global warming, not a natural increase in the suns output [47]. This technique uses mathematical models to show how the atmosphere would work if humans had not increased the CO_2 content of the atmosphere to 400 ppm. The method utilizes data from ancient paleoclimates and historical recent weather and shows that a hotter sun would heat the upper atmosphere and cool the lower atmosphere. The opposite has been observed in the post-industrial climate warming because the upper atmosphere has cooled and the lower atmosphere has warmed (e.g. Fig. 5.1). Climate change deniers have used the cooling of the upper atmosphere as

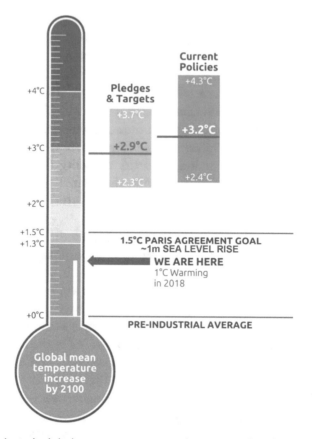

Fig. 5.15 Projected global mean temperature increase and sea level rise by 2100 based on Paris Agreement goal, pledges and targets of countries, and current policies of countries. The Antarctic ice melt and contribution to sea level rise (SLR) at 1 °C (2.12 °F) global average temperature increase is three times higher than what it was 20 years ago. In the past, over thousands of years, about 6 m of sea-level rise appears to have occurred at 1.5° C (3.2 °F) temperature increase, mainly from the west Antarctica and Greenland melting ice (see Figs. 5.5 and 5.9) (www.ipcc.ch › site › assets › uploads › 2018/02). A 2° C (4.2 °F) increase creates an even greater sea-level rise because Greenland seems to have had thresholds for near-complete melt near 2 °C (4.2 °F) and sea level rise of 7.2 m (24 ft) (https://en.wikipedia.org/wiki/Greenland_ice_sheet). Figure source is Climate Action Tracker, September 2019 update (https://climateactiontracker.org)

evidence for denying global warming, when it fact it indicates that humans are causing the global warming.

The fingerprinting statistical method has been applied to specific extreme weather events and other climate warming changes since the 1950s to show that humans are most likely to have contributed to, or caused these changes. When the 2003 European heat wave, killing thousands, had the highest temperatures since the introduction of weather instruments in 1851, fingerprinting indicated that humans were 75% to blame for the heat wave, and it was twice as likely to be human caused than natural (Begley, Newsweek, December 6, 2010). Using the same method, the 38 °C (100 °F) temperature of Siberia in 2020 is calculated to be 600 times more likely because of human-caused global warming. (Fig. 3.2) (https://www.worldweatherattribution.org/siberian-hea twave-of-2020-almost-impossible-without-climate-change/). Similar fingerprinting analyses also indicate that you need a large human contribution to climate warming to account for extreme precipitation and drought events, warming of the oceans and retreat of the Arctic sea ice (Figs. 4.9, 4.10, 5.2 and 5.12) [47].

With the above background we can attempt to separate the global warming caused by humans from natural climate variability. We can see that the earth has had a wide variation of atmospheric and oceanic temperatures associated with CO_2 in the atmosphere for millions of years (Fig. 5.5). Polar ice caps have come and gone and sea level has gone up and down because of natural causes such as volcanism, tilt of the earth's axis, deviation in the earth's orbit, shifting of the earths tectonic plates and changes in oceanic circulation (https://www.quantamagazine.org/how-earths-climate-changes-natura lly-and-why-things-are-different-now-20200721/?). The fundamental difference is that all of these natural processes changing CO_2 and temperature have taken thousands to millions of years to increase or decrease temperatures, except for a large meteorite or asteroid hitting the earth (Figs. 5.3 and 5.5) [1].

To differentiate natural from human-caused climate change, the rates of natural change for CO_2 in the atmosphere and ocean temperature can be compared to the post-industrial rates of change for these environmental characteristics. The best comparisons can be made with the most rapid rates of increase for natural CO_2 during the Pleistocene glacial to interglacial climates of the past 400,000 years, where the data are most accurate (Fig. 5.3). During Pleistocene times, CO_2 increased about 1 ppm per 100 years, whereas since the industrial revolution the increase is about 1 ppm per year or a human-caused rate that is 100 times faster than the most recent geologic rates of increase (Figs. 5.2 and 5.3).

Unfortunately the rate of human-caused CO_2 input is accelerating and from December 2018 until December 2019 the atmospheric CO_2 increase was 2.62 ppm or more than 200 times the natural late Pleistocene rates of increase (https://www.co2.earth/co2-acceleration). Similar natural versus post-industrial rates of change (i.e. 100 times faster) are observed in the sea surface temperatures. Between the last glacial to interglacial time of 10,000 years, the ocean temperature raised about 2.6 °C (4.7 °F) whereas the Bering Sea temperature raised about 2.6 °C (4.7 °F) during the past 100 years or an order of magnitude (10X) faster (Fig. 3.9) [3]. Unfortunately the rate of Bering Sea temperature rise, like surface temperature and CO_2, is also accelerating (Figs. 4.9, 5.1 and 5.2).

The only source for these unnatural rates of geologic change and global warming during the past 100 years is manmade carbon dioxide and methane. We have no sustained catastrophic volcanic activity, fast movements of tectonic plates, rapid changes in the tilt of the earth's axis or deviation in the earth's orbit, and no major changes in oceanic circulation. The sun has been dimming slightly for the last half-century while the Earth heats up, so global warming cannot be blamed on the sun. For a citation verifying statements of this paragraph and the most detailed and up to date analysis separating natural climate variability from human-caused global warming see this website: https://www.quantamagazine.org/how-earths-climate-changes-naturally-and-why-things-are-different-now-20200721/?

Our production of greenhouse gasses and the changes we see in CO_2 concentration in the atmosphere, temperatures of the air and ocean, and rise of sea level all correlate with population increase and human burning of fossil fuels (Figs. 5.1, 5.2, 5.4g–i, 5.7 and 5.8). Statistical analysis or fingerprinting indicates that the warming climate and increasing intensity of extreme weather events would not have happened without human influence [47]. We humans and other living beings are already experiencing the results of our global warming, but this is just the beginning of our global changes (Figs. 5.4, 5.7, 5.8, 5.12, 5.13 and 5.14). These changes could get much worse for us and the Earth if we keep going on as we are with fossil fuel use and pollution of our air, water and soil natural resources (see Chap. 6).

5.10 Message to Global Warming Deniers and Skeptics

It is anyone's right to deny human-caused global warming, even in the face of numerous measured scientific facts showing: (a) the obvious correlation between the rapid increase in population, energy use, and burning of fossil fuels that create CO_2 (e.g. Fig. 5.4g, i), (b) post-industrial consistent warming and rising sea level of the world's oceans (Figs. 5.3 and 5.7), (c) a pattern of increased severity of storms, floods, droughts, forest fires, species extinction and melting ice sheets at sea and on land (Figs. 4.10, 5.4f, 5.12 and 5.13), and (d) the overwhelming evidence that the present unnatural increase in CO_2 and rate of warming is much more rapid than natural geologic rates of changes (see Sect. 5.9).

If these examples of global warming do not get the attention of the denying politicians, perhaps the losses of life and hundreds of billions of dollars in costs may change their minds. To prevent further costs in lives and money, even climate change skeptics should realize that other natural global changes such as earthquakes and tsunamis should be prepared for (see Chap. 4 Sect. 4.13) in addition to other human caused global air pollution, water scarcity, and deforestation (see Chap. 6).

It should be thought provoking that USA after 2017 had a ruling populist political party that withdrew from the Paris Agreement. The withdrawal is especially significant since the USA is the largest single historical contributor of increased CO_2 and has emitted one third of the world's global warming CO_2 [29]. Why are there deniers and skeptics about climate science when other scientific research has revolutionized modern life with the development of electricity, telephones, airplanes, computers, the Internet and medical breakthroughs? Why has the USA been the world's biggest anomaly for climate change denial?

Although there is wide variation in numerous polls, prior to 2007 a majority (60–85%) of the United States population, like most of the Western world, believed that global warming is happening https://thebridge.agu.org/2013/06/05/global-warming-public-opinion-and-policy/). In 2007, Republican Senator McCain the candidate for USA president had climate change mitigation as part of his platform. However beginning in 2007, some irresponsible companies funded climate change denial, and USA Republican politicians as well as media gave equal weight to scientists denying climate change [29]. The media, because of concern about being biased against climate change deniers, became guilty of false equivalency. In other words the media provided equal coverage between the 2–3% of climate change denying scientists (many working for the fossil fuel industry) and the overwhelming majority of scientists that attribute climate change to human activities [41]. The USA media has even given 40% more climate change denial coverage than the world's other media.

The resulting confusion for the general population then caused: (A) nearly half of the USA population to deny or become skeptical of the scientifically accepted fact of global warming [25], and (B) the Republican Party to adapt a platform of denial, and (C) the USA to become the only country in the world to quit the Paris Agreement (https://www.ecowatch.com/fossil-fuel-industry-deceived-public-2641052019.html?rebelltitem=1#rebelltitem1). These changes are in spite of the fact that the newest analysis of 11,602 peer-reviewed articles on climate change published in 2019 shows there is 100% consensus among climate research scientists that global warming is human caused [41].

The good news now is there are fewer climate change deniers/skeptics in the USA population, unfortunately, because more and more people are personally affected by the devastating hurricanes, floods, wildfires, mudslides and bitter cold weather in 2017 and 2018 [26]. Psychologists suggest that only when people are personally affected do they change their strong opinions on political issues. The increasing number of people affected by the present day increase in severe extreme weather events, is proving the psychologists right.

Part of the problem contributing to skepticisim is that about 80 percent of media coverage of climate change has framed the subject in terms of disaster, according to a 2013 study by James Painter, of the Journalism Fellowship Program at the Reuters Institute for the Study of Journalism. Some particularly hysterical pieces have even proclaimed that changes will occur so rapidly that the world will be uninhabitable and we will all be dead in the next 10 years [13]. Because these irresponsible drastic predictions have not happened and also media sensationalism, this rhetoric leads to scientific disbelief and is no more responsible than the denial of climate change skeptics. In Chap. 7, I give a number of examples of governmental and private actions that are being taken to combat global change problems in spite of skepticism.

Why is climate change denial needed by the fossil fuel industries? The world is not suddenly going to stop using fossil fuels or requiring greater energy supplies (see Chap. 6, Sects. 6.2 and 6.3). The fossil fuel industry is not going to suddenly disappear. Probably within about half a century, most of the fossil fuels will be gone and we will need long-term renewable energy such as solar panels and windmills anyway. The faster countries develop renewable energy, the better to avoid an energy crisis later on. The world's prosperity and lifting people from poverty is based on the energy supply, and there is only the need to have a transition from the diminishing fossil fuels to a variety of sustainable energy supplies. The energy companies can continue to thrive and increase the number of jobs by transitioning to sustainable supplies from solar and wind power.

What is the cost of climate change denial other than the obvious loss of lives and property? Somehow the deniers have turned this issue into a fear of economic and job loss when the opposite is true. One of the biggest growth companies in USA recently has been Tesla, which makes electric cars. The

USA has created nearly a million jobs in sustainable energy, but denial is costing millions more. For example, when President Trump imposed a 30% tariff on solar panels, solar jobs fell for the first time in the history of the solar industry [29]. When the present USA government from 2017 to 2020 was stopping incentives for sustainable energy, China already had 3 million jobs in sustainable energy industries and was rapidly increasing this to 10 million jobs. Meanwhile, the USA was promoting coal use, which already had dropped from a million jobs to less than 100,000 in the past decades. And coal-fired plants for generating electricity are the least competitive type. At present, the cheapest electricity is generated by solar panels and windmills (3–4 cents per kwh), next cheapest by natural gas plants (5–6 cents per kwh) and coal plants kwh costs are about 50% more than this (Heal, 2017) (https://webstore.iea.org/worldenergy-outlook-2020).

Climate change denial has resulted in the populist USA federal government from 2017 to 2020 to subsidize the least economic and most polluting fossil fuels. Coal use for generating electricity is the least economic source and causes economic costs that do not account for health care, environmental damage, and disruption of activities in cities like Delhi India and Beijing China. As Fareed Zakaria has said on GPS, China is investing in a twenty-first century future and USA is investing in a twentieth century past and which country do you think will win in the future. The only people who will win in the USA are the few billionaire coalmine owners, many of who are going bankrupt anyway. If we want to cut government budgets, why subsidize a dying coal industry, energy inefficient ethanol from corn, and fossil fuel industries. Even if you deny climate change, it does not make any sense to promote an uneconomic future for USA.

The politicians and climate deniers need to stop demonizing government and recognize its crucial role in doing the most important thing that markets do not do, which is prioritizing and sustaining the common good [39]. In these times of the coronavirus pandemic, it is obvious that government scientists and agencies are necessary for the common good. The US Government created the Internet that has revolutionized the twenty-first century. Hardly any of the major technological developments of the twentieth century were produced by the private sector working alone. Entrepreneurs such as Thomas Edison and George Westinghouse developed electricity, but it took the federal government to build the delivery systems that brought electricity to most of Americans. The same is true of telephone service. The federal government, starting with President Dwight Eisenhower, was needed to build the USA interstate highway system.

Government action is needed immediately because global warming is happening already as shown by the multiple patterns of change described in this chapter. These changes affect every aspect of human lives and the environment, even including the quality of wine and skiing [38]. For example the annual maximum snow mass and snow season have decreased by 41% and 34 days on average over the western United States since 1980 [59].

Global warming also brings seemingly less sensational environmental disturbances, such as increasing variability in rainfall, earlier snowmelt and later freezing of permafrost. These actually are no less dangerous. Collectively, the effects of human-caused global change are causing massive loss of life, environmental and economic consequences globally, such as the loss of valuable topsoil, invasions of non-native plant and animal species, and increasing rates of disease [13]. The bottom line is that humans must adapt to climate warming no matter what caused it. Even if climate change deniers/skeptics are correct that climate warming of the past 100 years is natural, there is a proven link that more atmospheric CO_2 increases global temperatures, so why make any natural warming worse by adding more human produced CO_2.

Certainly the coronavirus pandemic of 2020 proves how the human race is globally linked and the critical need for scientific data. All of us, including climate change deniers and skeptics, need to take this in and realize that the coronavirus lesson is crucial for our global warming and all other global change problems. The coronavirus has spread almost instantaneously in time compared to global change problems. However, the slow spread of global changes is an even greater threat to the human race and should not be cast aside by fake news and unfounded scientific claims. Just as our best health science is required to solve the coronavirus pandemic, we need the same rigorous natural science to solve global change problems.

References

1. Alvarez LW, Alvarez W, Asaro F, Michel HV (1980) Extraterrestrial cause for the cretaceous-tertiary extinction. Science 208:1095–1108
2. Bednarsek N, Feely RA, Beck MW, Alin SR, Siedlecki SA, Calosi P, Norton EL, Saenger C, Strus J, Greeley D, Nezlin NP (2020) Exoskeleton dissolution with mechanoreceptor damage in larval Dungeness crab related to severity of present-day ocean acidification vertical gradients. Sci Total Environ 716. https://doi.org/10.1016/j.scitotenv.2020.136610
3. Bereiter B, Shackleton S, Baggenstos D, Kawamura K, Severinghaus J (2018) Mean global ocean temperatures during the last glacial transition. Nature 553:39–44. https://doi.org/10.1038/nature25152

4. Ceballos G, Ehrlich PR, Rodolfo D (2017) Biological annihilation via the ongoing sixth mass extinction signaled by vertebrate population losses and declines. Proc Natl Acad Sci Am 114:E6089–E6096. https://doi.org/10.1073/pnas.1704949114

5. Cheng L, Abraham J, Hausfather Z, Trenberth KE (2019) How fast are the oceans warming. Science 363:128–129. https://doi.org/10.1126/science.aav7619

6. Cramer BS, Miller G, Barrett PJ, Wright JD (2011) Late Cretaceous-Neogene trends in deep ocean temperature and continental ice volume: reconciling records of benthic foraminiferal geochemistry (d18O and Mg/Ca) with sea level history. J Geophys Res 116:C12023. https://doi.org/10.1029/2011JC007255,2011JGR

7. DeConto R, Pollard D (2016) Contribution of Antarctica to past and future sea-level rise. Nature 531:591–597. https://doi.org/10.1038/nature17145

8. Eagan D (2017) The life and death of the Great Lakes. Norton and Company, New York, p 364

9. Fasullo JT, Otto-Bliesner BL, Stevenson S (2018) ENSO's changing influence on temperature, precipitation, and wildfire in a warming climate. Geophys Res Lett 45:9216–9225

10. Francis JA, Vavrus SJ (2012) Evidence linking Arctic amplification to extreme weather in mid-latitudes. Geophys Res Lett 39:L06801. https://doi.org/10.1029/2012GL051000

11. Galeotti S, DeConto R, Naish T, Stocchi P, Florindo F, Pagani M, Barrett P, Bohaty SM, Lanci L, Pollard D, Sandroni S (2016) Antarctic ice sheet variability across the Eocene-Oligocene boundary climate transition, Science 352:76–80. https://doi.org/10.1126/science.aab0669

12. Goulson D (2019) Insect declines and why they matter, p 48. https://www.somersetwildlife.org/sites/default/files/2019-11/FULL%20AFI%20REPORT%20WEB1_1.pdf

13. Gronish E (2018) Doom-and-gloom scenarios on climate change won't solve our problem, Climate Change: Planet Under Pressure, Scientific American, iBooks. https://itunes.apple.com/WebObjects/MZStore.woa/wa/viewBook?id=163D2A3D07D0807D5C24F2DAB26C8E51

14. Gutjahr M, Ridgwell A, Sexton PF, Anagnostou E, Pearson PN, Pälike H, Norris RD, Thomas E, Foster GL (2017) Very large release of mostly volcanic carbon during the Palaeocene–Eocene thermal maximum. Nature 548. https://doi.org/10.1038/nature23646

15. Hansen J, Satoa M, Ruedy R, Schmidt GA, Lob K, Persin A (2018) Global temperature in 2017. http://www.columbia.edu/~jeh1/mailings/2019/20190206_Temperature2018.pdf

16. Hayhow DB, Ausden MA, Bradbury RB, Burnell D, Copeland AI, Crick HQP, Eaton MA, Frost T, Grice PV, Hall C, Harris, SJ, Morecroft MD, Noble DG, Pearce-Higgins JW, Watts O, and Williams JM (2017) The state of the UK's

birds 2017, The RSPB, BTO, WWT, DAERA, JNCC, NE and NRW, Sandy, Bedfordshire. 29 p

17. Huang J, Fu C (2017) Future looks drier as drylands continue to expand. EOS 98. https://doi.org/10.1029/2018EO086451

18. IPBES (2019) United Nations intergovernmental science-policy platform on biodiversity and ecosystem services, Global Assessment Summary for Policymakers. www.ipbes.net/news/ipbes-global-assessment-summary-policy makers-pdf

19. IPCC (2013) Climate Change 2013, The physical science basis. contribution of Working Group I to the fifth assessment *report* of the intergovernmental panel on climate change [Stocker TF, Qin D, Plattner G-K, Tignor M, SK, and others]

20. IPCC (2019) Special report on the ocean and cryosphere in a changing climate. https://www.ipcc.chsrocc

21. Kornei K (2019) 600 years of grape harvests document 20th century climate change. EOS 100. https://doi.org/10.1029/2019EO134355

22. Kunkel KE, Karl TR, Brooks H, Kossin J, Lawrimore JH, Amdt D, Bosart L, Changnon D, Cutter SL, Doesken N, Emanuel K (2013) Monitoring and understanding changes in extreme storms: state of knowledge. Bull Am Meteorol Soc 499–514. https://doi.org/10.1175/BAMS-D-12-00066.1

23. Kunzig R, Broecker W (2009) Fixing climate. CPI Bookmarque Ltd., Croydon, England, p 288

24. Labbe' T,Pfister C, Hnnimann S, Rousseau D, Franke J, Bois B (2019) The longest homogeneous series of grape harvest dates, Beaune 1354–2018, and its significance for the understanding of past and present climate. Climate of the Past 15:1485–1501. https://doi.org/10.5194/cp-15-1485-2019

25. Leiserowitz A, Maibach E, Rosenthal S, Kotcher J, Ballew M, Goldberg M, Gustafson A (2018) Climate change in the American mind: 2018, Yale University and George Mason University. Yale Program on Climate Change Communication, New Haven, CT, p 51

26. Leiserowitz A, Maibach E, Rosenthal S, Kotcher J, Bergquist P, Ballew M, Goldberg M, Gustafson A (2019) Climate change in the American mind: 2019, Yale University and George Mason University. Yale Program on Climate Change Communication, New Haven, CT, p 71. https://doi.org/10.17605/OSF.IO/CJ2NS

27. Masters J (2014) The jet stream is getting weird. Sci Am 311:68–75

28. McClintock J (2012) Lost Antarctica: adventures in a disappearing land. Martins Press, St, p 256

29. McKibben W (2018) Life on a shrinking planet. The New Yorker 47–55

30. Michon S, Lindsey R (2018) Which emits more carbon dioxide: volcanoes or human activities? NOAA Climate.gov. https://www.climate.gov/print/814945, p 7

31. Miller KG, Wright JD, Browning JV, Kulpecz A, Kominz M, Naish TR, Cramer BS, Rosenthal Y, Peltier WR, Sosdian S (2012) High tide of the warm Pliocene:

implications of global sea level for Antarctic deglaciation. Geology 40:407–410. https://doi.org/10.1130/G32869.1

32. Miller SM, Taylor MA, Watts JD (2018) Understanding high-latitude methane in a warming climate. EOS 99. https://doi.org/10.1029/2018EO091947

33. Milly PC, Dunne KA (2020) Colorado River flow dwindles as warming-driven loss of reflective snow energizes evaporation. Science https://doi.org/10.1126/science.aay9187

34. Moskvitch K (2014) Mysterious Siberian crater attributed to methane. Nature. https://doi.org/10.1038/nature.2014.15649

35. Nerem RS, Chambers DP, Choe C, Mitchum GT (2010) Estimating mean sea level change from the TOPEX and JASON altimeter missions. Mar Geodesy 33(sup1):435–446. https://doi.org/10.1080/01490419.2010.491031

36. Nerem RS, Beckley BD, Fasullo JT, Hamlington BD, Masters D, Mitchum GT (2018) Climate-change–driven accelerated sea-level rise detected in the altimeter era. Proc Natl Acad Sci 115:2022–2025. https://doi.org/10.1073/pnas.1717312115

37. Netflix (2020a) David Attenborough: a life on our planet. https://www.neflix.ctom/es-en/title/80216393

38. Nicholas KA (2018) Will we still enjoy pinot noir? Climate change: planet under pressure. Scientific American, iBooks. https://itunes.apple.com/WebObjects/MZStore.woa/wa/viewBook?id=163D2A3D07D0807D5C24F2DAB26C8E51

39. Oreskies N (2018) How to break the climate deadlock, Scientific American. Climate change: planet under pressure, iBooks. https://itunes.apple.com/WebObjects/MZStore.woa/wa/viewBook?id=163D2A3D07D0807D5C24F2DAB26C8E51

40. Osborne E, Richter-Menge J, Jeffries M (eds) (2018) Arctic Report Card 2018. https://www.arctic.noaa.gov/Report-Card

41. Powell J (2019) Scientists reach 100% concensus on Anthropogenic global warming. Bull Sci Technol Soc 39:2. https://doi.org/10.1177/0270467619886266

42. Pross J, Contreras L, Bijl PK, Greenwood DR, Bohaty SM, Schouten S, Bendle JA, Röhl U, Tauxe L, Raine JI, Huck CE, van de Flierdt T, Jamieson SSR, Stickley CE, van de Schootbrugge B, Escutia C, Brinkhuis H, IODP Expedition 318 Scientists (2012) Persistent near-tropical warmth on the Antarctic continent during the early Eocene epoch. Nature 488:73–77. https://doi.org/10.1038/nature11300

43. Reuters (2019) Scientists shocked by Arctic permafrost thawing 70 years sooner than predicted, The Guardian. ISSN 0261-3077

44. Rignot E, Mouginot J, Scheuchl B, van den Broeke M, van Wessem MJ, Morlighem M (2019) Four decades of Antarctic ice sheet mass balance from 1979–2017. Proc Natl Acad Sci 116:1095–1103. https://doi.org/10.1073/pnas.1812883116

45. Rosenberg KV, Dokter AM, Blancher PJ, Sauer JR, Smith AC, Smith PA, Stanton JC, Panjabi A, Helft L, Parr M, Marra PP (2019) Decline of North American avifauna, Science 366:120–124. https://doi.org/10.1126/science.aaw 1313

46. Sanchez-Bayo F, Wyckhuys KAG (2019) Worldwide decline of the entomo-fauna: a review of its drivers. Biol Cons 232:8–27

47. Santer BD, Wigley TML (2007) Progress in detection and attribution research, Program for Climate Model Diagnosis and Intercomparison, Lawrence Livermore National Laboratory. Livermore, CA 94550, USA. in Climate Change Science and Policy, p 28. http://www.image.ucar.edu/idag/Papers/PapersIDA Gsubtask1.4/Schneider_bookchapDnA.pdf

48. Schuur T (2016) The permafrost prediction. Sci Am 315:56–61

49. Seviour WJM (2017) Weakening and shift of the Arctic stratospheric polar vortex: Internal variability or forced response? Geophys Res Lett 44:3365–3373. https://doi.org/10.1002/2017GLOU3071

50. Shen Q, Wang H, Shum CK, Jiang L, Hsu HT, Dong J (2018) Recent high-resolution ice velocity maps reveal increased mass loss in Wilkes Land, east Antarctica. Nat Sci Rep 8:1–7. https://doi.org/10.1038/s41598-018-22765-0

51. Shakhova N, Semiletov I, Salyuk A, Kosmach D (2008) Anomalies of methane in the atmosphere over the East Siberian shelf: Is there any sign of methane leakage from shallow shelf hydrates?. Geophys Res Abstracts 10, EGU2008-A-01526

52. Slater T, Lawrence IR, Otosaka IN, Shepherd A, Gourmelen N, Jakob L, Tepes P, Gilbert L (2020) Review article: earth's ice imbalance, The Cryosphere Discuss. https://doi.org/10.5194/tc-2020-232, in review, 2020

53. Tanski G, Wagner D, Knoblach C, Sachs T, Lantuit H (2019) Rapid CO_2 release from eroding permafrost in seawater. Geophys Res 46:11244–11252. https://doi.org/10.1029/2019GL084303

54. Tripati AK, Roberts CD, Eagle RA (2009) Coupling of CO_2 and ice sheet stability over major climate transitions of the last 20 million years. Science 326:1394–1397. https://doi.org/10.1126/science.1178296

55. Watson CS, White NJ, Church JA, King MA, Burgette RJ, Legresy B (2015) Unabated global mean sea-level rise over the satellite altimeter era. Nat Clim Chang 5:565–568

56. Wiens JJ (2016) Climate-related local extinctions are already widespread among plant and animal species. PLOS Biol 14:e2001104. https://doi.org/10.1371/jou rnal.pbio.2001104

57. Wilson DJ, Bertram RA, Needham EF, van de Flierdt T, Welsh KJ, McKay RM, Mazumder A, Riesselman CR, Jimenez-Espejo FJ, Escutia C (2018) Ice loss from the East Antarctic Ice Sheet during late Pleistocene interglacials. Nat Lett 561:383–392. https://doi.org/10.1038/s41586-018-0501-8

58. Wing OEJ, Bates PD, Smith AM, Sampson CC, Johnson KA, Fargione J, Morefield P (2018) Estimates of present and future flood risk in the conterminous

United States. Environ Res Lett 13(7):034023. https://doi.org/10.1088/1748-9326/aaac65

59. Zeng X, Boxton P, Dawson N (2018) Snowpack change from 1982 to 2016 over conterminous United States. Geophys Res Lett 45:12940–12947. https://doi.org/10.1029/2018GL079621

60. Zhu Z, Piao S, Myneni RB, Huang M, Zeng Z, Canadell JG, Ciais P, Sitch S, Friedlingstein P, Arneth A, Cao C (2016) Greening of the Earth and its drivers. Nat Clim Change 6:791–795

61. Corell R (2005) Arctic climate impact assessment. Bulletin of the American Meteorological Society 86:860–861

62. Ripple WJ, Wolf C, Newsome TM, Galetti M, Alamgir M, Crist E, Mahmoud MI, Laurance WF, (2017) World Scientists' Warning to Humanity: A Second Notice. BioScience 67(12):1026–1028

63. Cheng L, Trenberth KE, Fausullo J, Boyer T, Abraham J, Zhu J (2017) Improved estimates of ocean heat content from 1960 to 2015. Science Advances 3:e160145. https://doi.org/10.1126/sciadv.1601545

6

Other Global Changes

6.1 The Importance of Combined Global Changes

The culmination of my experience is realizing the importance of the multiple global changes that the human race faces within the next century (e.g. Netflix [31]) (see the acceleration of changes in Figs. 5.4, 6.1 and 6.2). I have combined the knowledge of my geologist's journey with new data of other sources to summarize some of the most important global changes that humans have caused and are facing (e.g. Netflix [30, 32]). In addition, my unusual global perspective from terrestrial and ocean studies while working in government, academia and industry for over 60 years has given me some perspective to suggest solutions to provide for a sustainable Earth.

In my book introduction, I originally pointed out that severe human-caused global changes are rapidly affecting our future for a sustainable Earth. As described throughout the text, at present rates, these main global changes during the next century include: (1) global warming of the climate and increased air pollution (Figs. 5.1, 5.2, 5.3, 5.4, 5.5, 5.6, 5.7, 5.8, 5.9, 5.10, 5.11, 5.12, 5.13, 5.14 and 5.15); a potential 50% loss of potable water (Figs. 5.4b, 6.1 Water use) and most farmland because of pollution and lack of conservation practices (5.2 Domesticated land, Terrestrial biosphere degradation); increased intensity of floods (Fig. 5.12), drought, desertification (Fig. 5.4e), and wildfires (Fig. 5.13); (3) use of most of our petroleum

© The Author(s), under exclusive license to Springer Nature
Switzerland AG 2021
C. H. Nelson, *Witness To A Changing Earth*,
https://doi.org/10.1007/978-3-030-71811-4_6

Socio-economic trends

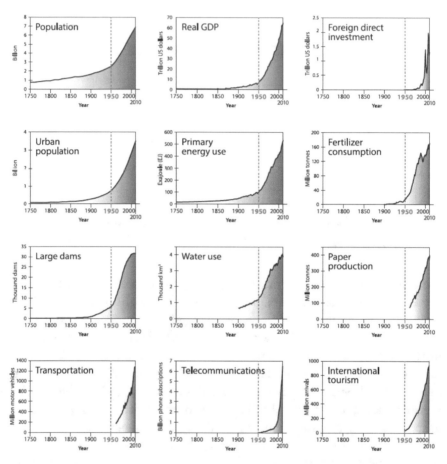

Fig. 6.1 Acceleration of socio-economic trends that parallel the rapid population growth since 1950. These trends show the use and depletion of the Earth's resources that support the human population. See [1] for the data source for each box in the figure

resources (Fig. 6.1 Primary energy use); and (4) loss of the world's wild fisheries (Figs. 5.4c, 6.2 Marine fish capture), coral reefs and mangrove forests (5.2 Tropical forest loss). At the same time as we deplete these resources that sustain the human population, the world's population will increase by about 50% (e.g. Figs. 5.4i and 6.1 Population) (https://en.wikipedia.org/wiki/Projections_of_population_growth). The combination of these global changes and increased population over a century does not bode well for humans or other biological life. We already are seeing a 60% decrease in abundance

Earth system trends

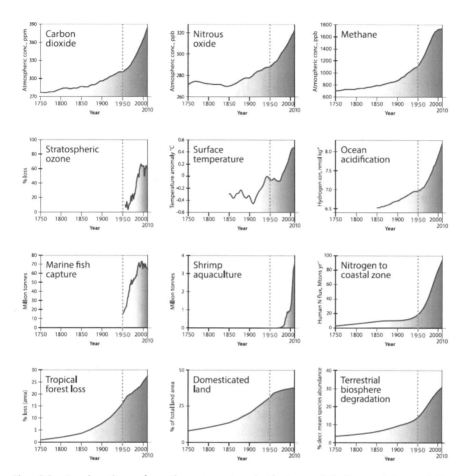

Fig. 6.2 Acceleration of earth system trends that parallel the rapid population growth since 1950. These trends show the pollution and depletion of the Earth's resources that support the human population. See [1] for the data source for each box in the figure

of vertebrate species (Fig. 5.4f). Humans also are vertebrates and can face extinction unless we develop the social and political will to reverse these human-caused global changes.

To reduce the human-caused global changes and sustain as well as improve conditions for humans, the United Nations has developed a set of 17 Sustainable Development Goals (SDGs) to be achieved by the year 2030 (https://sustainabledevelopment.un.org/?menu=1300). The UN blueprint, for peace and prosperity for people and the planet, focuses on 17 topics such as economic, educational, equality, energy and environmental goals. In the following sections, I will address some of these UN topics related to global warming, energy and environmental goals of clean air and safe drinking water for life on land and in the ocean.

6.2   Energy and Global Warming

Human energy use has grown exponentially since the industrial revolution and has rapidly increased the CO_2 and CH_4 content of the atmosphere and led to global warming (Figs 5.2, 5.3, 5.4g, h, 5.6 and 6.1 Primary energy use, 5.2 Methane). The CO_2 and CH_4 that has deposited in the earth over hundreds of millions of years is now being released to the atmosphere in a century or two and causing global warming. However, we cannot go cold turkey and suddenly stop burning fossil fuel that provides the energy for the world's economy and sustains the human population. A more rational approach is to buy time with the Earth's remaining and limited fossil fuel resources so that the world's infrastructure can be revised to provide renewable energy supplies from hydroelectric, wind, solar power and perhaps nuclear power. Nuclear power however, requires that proper studies be done for geological hazards at sites for nuclear power plants and at locations for long-term storage of nuclear wastes.

The use of hybrid and electric cars can help this transition to sustainable energy and provide conservation of the existing fossil fuels. Conservation of energy of all types is the most rapid way to cut the use of fossil fuels. Europe has shown, that the developed world with a high standard of living can use half energy compared to the per capita use of energy in the United States (https://www.google.com/search?client=firefox-b-d&q=compar ison+of+USA+and+european+energy). In the USA there is less use of sustainable energy and more use of energy inefficient vehicles, electrical appliances and electrical energy in buildings. Energy conservation is not a hardship because it is a win situation where when energy is conserved, the costs for private users and businesses are reduced.

The worst approach to energy use is uncontrolled development such as utilizing huge coal strip mines and permitting the wastes to be dumped into streams. There will have to be some use of coal as the transition to sustainable energy takes place. However, the true environmental cost of these coalmines must be considered and provide for reclamation of the mined areas. Also mines must be regulated to prohibit the dumping of mining waste and leakage of coal ash into surface and ground water. These same suggestions apply to mineral mines for metals as well, where wastes should not be dumped into drainages where toxic heavy metals (e.g. lead, cadmium, arsenic) can pollute for extensive distances down streams and rivers (e.g. Nelson and Lamothe [2]). For years the uncontrolled development of coal and mineral mines has destroyed huge areas of the Earth because the total cost of having non-polluting mines has not been included in the price of coal and metals.

The controversial use of fracking to produce petroleum and natural gas at least has been reducing the use of coal fired electrical plants in the United States. Again there can be a rational approach to the use of fracking, which has fostered the positive replacement of coal-fired plants and energy independence for the United States. As usual, the main problem has been the rapid uncontrolled development of fracking without the proper use of environmental assessments before fracking has been undertaken. Each area where fracking is undertaken has its own environmental setting such as susceptibility of groundwater pollution and earthquakes from injection of chemicals and water for fracking. If the proper studies are done to assess whether fracking will contaminate groundwater or create earthquakes, fracking processes can be utilized in the areas that are safe. With proper environmental assessment it is not necessary to completely halt fracking for petroleum, but to allow it only in safe areas.

A friend of mine is one of the world's leading experts on fractures in rocks and I discussed with him the problems of water injection from fracking that has caused a significant increase in earthquakes in areas such as Oklahoma. He mentioned there was a previous history of water injection for petroleum recovery at many of the locations where water and chemical injection for fracking produced numerous new and stronger earthquakes reservoirs. Consequently, if areas are known to have had significant water injection prior to fracking, careful environmental assessment should be done to determine whether additional water injection for fracking could produce earthquakes. Environmental assessment studies should be done prior to drilling and fracking in any location, and then this information needs to be utilized by government agencies to prohibit fracking for petroleum where there is a risk for creating earthquakes or contaminating the ground water.

There also can be a rational approach to offshore drilling for petroleum, so that the proper environmental assessments, regulations and procedures for development are followed. The offshore drilling for petroleum has had a good history for safety such as offshore Brazil where no major petroleum disasters have occurred in large areas of extensive deep-water drilling. On the other hand, the catastrophic Horizon deepwater drilling disaster in the Gulf of

Mexico in 2010 was entirely preventable if proper procedures and regulations had been in place (see Chap. 3 Sect. 3.8). British Petroleum Company (BP) had been one of the most environmentally responsible companies by acknowledging climate change and instituting major conservation practices that saved them billions of dollars in energy costs for the company. However in the twenty-first century, BP began to cut corners to increase profits, and evidence of this cost-saving was provided by several refinery fires and lack of maintenance on the Alaska pipeline. In 2010 at the Horizon drilling platform they rushed to complete the oil well and used sub-standard cement and a faulty blowout preventer, which resulted in the largest oil spill in history.

The petroleum industry has a strong vested interest and takes extensive precautions to prevent such oil spill disasters, because something like the Deepwater Horizon oil spill cost BP nearly 70 billion dollars (https://en.wikipedia.org/wiki/Deepwater_Horizon_oil_spill). However, there are other measures that should be undertaken by petroleum companies to help prevent global change. One of these is the flaring or burning of methane (CH_4) at oil platforms on land and sea. Flaring creates unnecessary pollution and use of petroleum resources, just to save the companies money. Another measure is that CH_4 created at fracking sites should not be released to the atmosphere and it should be contained for use as a resource and not wasted. Methane is a more potent global warming gas than CO_2 and thus all steps should be taken to reduce the influx of it into the atmosphere (Figs. 5.6 and 6.2 Methane).

The bottom line is that the burning of fossil fuels needs to be reduced as rapidly as possible while there is a transition to renewable energy. On the positive side, renewable energy is an area where progress is being made for a more sustainable Earth. Several countries are now planning to switch completely to electric vehicles within a few decades. Most important of these are China and India, which have the world's largest populations and have recently been creating the most increase of CO_2 and use of fossil fuels (Fig. 6.3). These countries, like USA, also are beginning to eliminate coal-fired power plants that create the most CO_2 and other environmental pollutants. This is being helped by the large new discoveries of natural gas (e.g. Mozambique, USA fracking), which has made burning the less polluting natural gas, cheaper than coal to produce electricity [27].

Most promising is the use of sustainable wind and solar power that use no fossil fuels and in parts of the world produce electricity for as little as half of the cost of natural gas-fired power plants (https://webstore.iea.org/worldenergy-outlook-2020). As Heal [27] points out in his book on environmental economics, when using solar panels in the southern USA, electricity is produced at a cost of approximately four cents per kilowatt-hour (kwh),

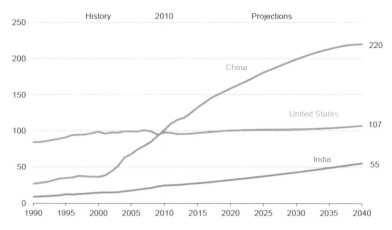

Fig. 6.3 Previous and projected energy consumption in quadrillion Btu for China, United States and India from 1990 to 2040. Note that China's energy use has more than tripled since the year 2000 and will be more than double that of the United States by 2040. Figure source is Sieminski [28]

and in the Middle East it costs approximately three cents (https://webstore. iea.org/worldenergy-outlook-2020). This is compared to the newest and most efficient natural gas power plants where electricity is produced for five to six cents per kilowatt hour (kwh), when natural gas prices are low. It costs even more to produce electricity in coal and oil fired plants. Windmills generate the cheapest power in USA at 3.5 cents per kwh. This is why some countries like Scotland generate 100%, Denmark nearly 40% and Spain 22% of their electricity using wind power. Countries such as Isreal and Spain already require that all new houses use solar panels to heat water. Personally, I have installed solar panels for water heating several years ago, cut my fuel oil use in half per year and the fuel savings have already paid for the solar panels, with free water heating for the future.

The effect of renewable energy production is already showing up in projections of the oil and gas industry. Previously this industry was concerned about peak oil production, which Dr. Hubbard of the USGS in 1940, originally predicted would take place at about 1970 in the USA. After that the production of oil and gas would decline and eventually be depleted. This projection was close to correct until the twenty-first century when new technologies resulted in major oil and gas discoveries in the offshore deep water in the Gulf of Mexico, and in fracking for petroleum.

Worldwide, not just in the USA, peak oil originally was predicted to occur prior to the middle of the twenty-first century, but also began to shift later with the new technologies and increased petroleum discoveries. However, new discoveries of petroleum have begun to drop significantly in the last few years and when worldwide peak oil may occur is now under debate. This debate also has now changed because of the significant increase in renewable energy supplies and use of hybrid and electric vehicles. The petroleum industry is now debating whether peak oil may occur or whether peak oil demand may occur. In other words because of increasing renewable energy, petroleum demand may decrease enough so that peak oil production may never occur. The bottom line is that the energy future looks much more positive because of renewable energy use and hybrid/electric vehicles. Unfortunately, reduced energy use is taking place too slowly to meet the Paris Climate Agreement goals (Figs. 5.15 and 6.3).

6.3   The Energy Future

The future of energy use is a fundamental key to the future of a sustainable planet. On the one hand, the burning of fossil fuels has created global warming with the worldwide pattern of increased intensity of extreme weather tragedies (e.g. in 2017 alone thousands of deaths and 100 s of billions of dollars of damage in Asia and North America). On the other hand, increasing energy use of the industrial revolution has transformed the living standards of over 1.3 billion people of the western world and is the key to bettering the living standards of the other more than 6.1 billion people in the rest of the world (Fig. 6.1 Population, Real GDP) (Medlock III, AAPG, [29]).

Energy supply in the future will be the key to sustaining the developed world and lifting as many people as possible out of poverty (Tinker, AAPG, [29]). Unfortunately by denying global warming, providing funding for political campaigns, and misinformation to the media, the fossil fuel industry has misused this philosophy of supporting the global economies and the misguided goal of the next quarters profits to encourage the burning of fossil fuels as rapidly as possible. As a geologist taking a long view of the Earth over billions of years, this denial of global warming and pollution of the air, water and land seems very shortsighted. First of all the folks in the fossil fuel industry have families that need to live on a sustainable earth and second of all, the more the industry encourages the burning of fossil fuels, the shorter time their companies will have a viable industry.

In terms of the energy future for the fossil fuel industry, predicting the peak oil time has been an important, but complicated debate, because of changing technologies and population growth, rapidly increasing renewable energy development, and global political instability (e.g. wars, sanctions, mismanagement) in some major oil producing regions. Two of the most recent estimates in 2016 and 2017 by British Petroleum and Shell Oil companies suggest that peak oil will occur in about 50 years (Yielding, AAPG [29]). British Petroleum Company predicts that the main burning of fossil fuels will zero out by about 2100 and the remaining use will be for aviation fuel and non-combustible products like plastics. Shell Oil Company however, predicts that in 2100 total fossil fuel use for combustion will still remain at about 20%.

It is troubling that for the next 50 years, both companies predict that demand for fossil fuel will nearly double and use will remain at about 30% each for coal, oil and gas (Yielding, AAPG [29]). As a result of this, CO_2 is predicted to increase by 25% in 2030, and for the worst-case scenario, CO_2 concentration is expected to be 600 ppm by 2060 or by 2100 (Koonan, AAPG [29]) (biowesleyan.wordpress.com). Eighty percent of this increased demand is from Asia, (mainly China and India) and 50% of the present worlds coal combustion is in China (Fig. 6.3) (Snell, AAPG [29]). Of course these CO_2 amounts are not in line with the required commitments from countries to meet the Paris Agreement goals. Fortunately in both North America, which peaked in 2007, and Europe, energy demand has remained static during the twenty-first century and should decline with the increasing use of renewable energy (Fig. 6.3) (Ausubel, AAPG [29]).

There has been recent worldwide progress in the use of renewable energy and in the conversion of coal-fired power plants to gas-fired plants, with the hope that all coal-fired plants will be eliminated by the year 2100. Scotland now obtains 90% of it electricity from sustainable energy and Spain already obtains nearly 50% of its electricity from wind, hydro and solar power (https://en.wikipedia.org/wiki/Renewable_energy_in_Scotland) (renewableenergyworld.com). Unfortunately, United States and China only get electricity from about 12% of renewable energy sources. On the positive side, USA gets only 10% of its electricity from coal burning plants compared to China obtaining 80%, and USA sulfur emissions from coal burning peaked in 1970 (Ausubel, AAPG [29]). China is making an effort to reduce coal burning power plants and now produces two thirds of the solar panels and half of the wind turbines for the world. This effort not only benefits sustainable energy, but it results in a huge number of new jobs in China. There are three million jobs in the solar power industry in China now and this is expected to increase to ten million. The USA, compared to China has only one-tenth the number of jobs in the solar industry now. The USA has lost the potential of millions of new jobs because of its denial of global warming and it will take decades to reach the percent of renewable energy that now is used in Europe.

The USA also needs to evaluate not only the cost of job loss because of global warming denial, but in addition realize the true cost of its bio-fuel industry that results from political lobbying for farming subsidies and ethanol use. Both British Petroleum and Shell Oil project that up to a third of the future energy source in 2100 will be from bio-fuels (Yielding, AAPG [29]). However, presently the USA produces its bio-fuel from corn, which requires an area equivalent to the farmland of Iowa or Alabama (Ausubel, AAPG [29]). Utilizing corn for bio-fuel requires 1.2 units of energy to produce 1 unit of energy. Seven times more energy is expended to produce ethanol from corn than from sugar cane, two times more farmland is required and the cost is 15% more per unit of ethanol produced (https://www.biowesleyan. wordpress.com).In addition, the use of corn for cropland and loss of USA farmland for other grain has driven up the worldwide cost of grain, which has resulted in increased malnutrition in the developing countries. It is much more prudent to produce bio-fuels from crops like sugar cane rather than corn.

Much of the information for this Sect. 5.3 comes from a summary of the energy future that was presented at The American Association of Petroleum Geologists (AAPG) 2017 forum: The Next 100 Years of Global Energy Use: Resources, Impacts and Economics. This series of power point talks can be obtained from the following 2017 AAPG Search and Discovery Articles: Tinker #70,268; Koonin #70,269; Yielding #70,270; Snell #70,271; Ausubel #70,272; Medlock III #70,273.

6.4   Air Pollution

The World Health Organization (WHO) calls air pollution the invisible killer, as it can be difficult to trace, yet is responsible for 36% of lung cancer, 35% of pulmonary disease, and 27% of heart disease fatalities each year. The Lancet Commission on Pollution and Health in 2017 reported that pollution is responsible for 9 million global deaths each year with 72% of these resulting from air pollution and causing 25% of deaths in low and middle-income countries. Examples of this problem occurred in New Delhi India in 2017 and 2019 when the air pollution index reached 1000, whereas WHO indicates that greater than 25 is unsafe (https://en.wikipedia.org/wiki/Air_pol lution_in_Delhi). Breathing the New Delhi air was equivalent to smoking 44 cigarettes a day. Similar conditions occur in some Chinese cities and I recall during my visit that the polluted gray skies often obscured the sun.

For a sustainable Earth, and especially a sustainable human population, we need clean air. In my lifetime I have observed effects of the increasing air pollution, such as the epidemic increase of asthma and allergies. Back in the 1950s a few friends and family had what was called hay fever and only rarely asthma. Now allergies and asthma are common and increasingly so with young children. Fortunately, air pollution and related diseases may diminish in the future, particularly because China and India are beginning to institute pollution controls and have plans to eliminate all coal-fired power plants and to convert to all electric cars within a few decades.

There are many global changes that are affecting air quality and some such as burning fossil fuels, I have already mentioned. The biggest effect of burning fossil fuels is introducing CO_2 into the air, which then is a main contributor to global warming, as explained in Chap. 4 (Figs. 5.1, 5.2, 5.4g, h and 6.2 Carbon dioxide). Burning coal also introduces a host of other atmospheric pollutants such as soot, heavy metals, particularly mercury, and sulfides. Sulfides turn into acid rain which acidifies aquatic environments throughout North America and other continents in the northern hemisphere where the world's population is concentrated (Britannica.com). The acidification of water also results in the formation of methyl mercury, which can cause bioaccumulation in other biota including humans.

Burning petroleum products causes the same air pollution as burning coal, although to a lesser degree for chemical pollution of heavy metals and sulfides. However, increased atmospheric pollution of nitrous oxide from high compression gasoline and diesel engines is a less well-known effect of vehicles burning fossil fuels (Fig. 6.2 nitrous oxide). The nitrous oxide from these engine exhausts enters worldwide lake waters to increase the nitrogen content of the water. As previously described, Lake Tahoe again provides an example of a result of this atmospheric air-fall because its unpolluted nitrogen to phosphorous ratio should be 1–1, but at present this ratio is as much as 40–1 [3]. This causes unwanted biological productivity, such as algal growth and reduced clarity of the water. Additional productivity in Lake Tahoe also results from global warming, because the lake has warmed by 0.24 °C (0.5 °F) each year for the past four years. This is 14 times faster than the historic rate. As a result of the warmer surface waters, the lake does not mix water as deep as normal and more nitrates accumulate to grow algae.

Severe global air contamination also resulted from motor vehicles when companies added lead into gasoline. It spread throughout the global atmosphere and could be found in all atmospheric, water, and land environments on earth. Fortunately, this pollution problem provided one of the best examples of global cooperation, because leaded gasoline has been nearly eliminated and this global air pollution from lead has been mainly reduced. A similar success story is the Montreal Protocol, which significantly diminished use of per fluorocarbons (PDFs) (Fig. 5.4a). The escape into the atmosphere of PDFs from refrigerants and aerosol cans had caused a great reduction in the ozone layer over the Earth's poles. This has a number of effects, but for humans it allows more ultraviolet radiation, an increase in severe sunburns, and eventually more melanoma cancers. Melanoma has increased more than 1000% in the past few decades, although much of this can be attributed to people's previous lack of knowledge about sunburn and the creation of cancer many years afterwards.

There are other human-caused factors that contribute to both local and global pollution of the air. Chemical plants and petroleum refineries cause extensive environmental damage locally downwind from the plants and then globally because of the input of CO_2 and CH_4, which contribute to global warming. Again with regulations such as prohibiting gas flaring and escape of methane, these air pollutants can be reduced.

Another human-caused activity with implications for both local and global atmosphere is the rapid worldwide deforestation (Figs. 5.4e and 6.2 Tropical forest loss) and increased intensity and burned areas of forest fires (Fig. 5.13). The deforestation causes local effects such as soil erosion and rapid introduction of sediment to streams and rivers. This destroys the natural environment and eliminates the clear water necessary for trout and salmon to spawn. Less forest in tropical wet areas increases surface runoff and results in more flooding and human casualties (https://link.springer.com/article/10.1007/s10 640-014-9834-4).

An additional important global effect of the deforestation and forest fires is increasing the content of CO_2 in the air because the forests are gone and no longer absorb large amounts of CO_2 from the atmosphere (Fig. 5.13). Also the loss of forests results in decreased water vapor put into the atmosphere, which amplifies the affects of drought. The droughts from global warming cause stress in trees so that diseases and pine bark beetle infestions kill trees, which feed the intensity of forest fires. For example 12 million trees have died from infestations in California and since 2015 the number of dead trees has increased from 35,000 to 72,000 in the Lake Tahoe basin [3]. The increased intensity and area of forest fires from global warming, not only adds CO_2 to the atmosphere, but also causes significant air pollution from smoke and soot over wide areas, such as much of western North America in 2017, 2018 and 2020.

In the western USA in 2017, the smoke pollution from wildfires closed down major highways, airports, school systems, and outdoor cultural events (Fig. 1.9). This pales in comparison to the hundreds of lives lost, thousands of homes destroyed, disastrous mudslides and millions of square kilometers of areas burned in western USA in 2017, 2018 and 2020. I thought the 2017 fires were my worst experience with forest fires, but this was nothing compared with the fires in 2018 and 2020. In July 2018, our family went to Yosemite National Park and could not see a thing because of smoke from the Ferguson fire and then the park was closed down for several weeks. I took Spanish relatives to Crater Lake National Park and could not see the lake because of smoke from the numerous fires in southern Oregon. We camped on the edge of Diamond Lake and did not see the lake. We had a cabin in the Rogue River Valley, and for two months the valley had the worst air quality in USA. The only worst possible air quality was in an actual fire zone. For weeks in the Rogue Valley, children could not go outdoors.

I have been going to the Ashland, Oregon Shakespeare festival and Jacksonville, Oregon Britt Music festival for 60 years and for the first time in 2017 and 2018, outdoor concerts and plays had to be moved indoors. In 2018, I drove through smoke the entire way from central Oregon to San Francisco one time because of the Redding fire and the Mendocino fire, the largest fire in the history of California until 2020. For nearly a week I-5 the main north to south freeway in western North America was closed and the normal six-hour driving time between San Francisco and the California border became 13 h.

Unfortunately, the wildfires have been even worse in 2020. California recorded its first "gigafire": a single blaze that scorched more than 1 million acres and an area roughly the size of Connecticut had been burned by October (https://www.economist.com/graphic-detail/2020/10/14/this-is-the-worst-fire-season-the-american-west-has-ever-seen). Besides destroying hundreds of homes and killing dozens of people, the catastrophic fires have also put surviving residents' health at risk. Cities such as San Francisco, Portland and Seattle have suffered some of the most polluted air on the planet and people there are breathing greater amounts of particulate matter. A new study estimates that every additional microgram per cubic meter of daily PM2.5 (particulate matter less than 2.5 microns in diameter) exposure causes 0.69 additional deaths per million people aged 65 or over. Based on this calculation, the poor air quality from the 2020 wildfires will kill thousands of mainly elderly people, just in California (https://siepr.stanford.edu/research/publications/managing-growing-cost-wildfire).

The loss of life and huge costs of fires are related to global warming, which has caused increased hot, dry weather and a continuing decline in soil moisture (Fig. 5.1) (USA TODAY, Jan 8, 2019). Although there may not be more fires, the number of acres burned has doubled since 2000 and the fires now spread much faster and over much greater areas because of the increasingly lower soil moisture and more dead trees from pine bark beetle infestations caused by the warming climate (Fig. 5.13) (https://siepr.stanford.edu/research/publications/managing-growing-cost-wildfire). The economic costs also have increased dramatically and for California alone in 2018, the estimated cost just for insurers was 13 billion dollars. Some sources even estimated that all direct and environmental total costs of fires in 2018 were as much as 400 billion dollars or 2% of total USA GDP (https://www.accuweather.com). Even worse, the California Camp fire, which I could see making the sky red for miles while driving along I-5 freeway, killed 86 people, the worst in California history.

NOAA's annual billion-dollar disasters summary puts the 2018 forest fire and other increased extreme weather events in another perspective. The agency keeps track of weather disasters in the United States that account for more than $1 billion in direct losses. Since the tally began in 1980, the average number of billion-dollar disasters per year has been 6.2. In 2018, however, there were 14 events topping $1 billion. Western wildfires caused the majority of the losses, costing $73 billion of the $91 billion total. The worlds top three costliest natural disasters in 2018 all took place in the United States, and the Camp Fire was the worst costing 16 billion dollars (USA Today, Jan 8, 2019). In total, wildfires, hurricanes and other extreme weather cost the USA 247 lives and 91 billion dollars in 2018.

As mentioned previously, forest fires also cause health hazards because they introduce large amounts of soot particles (Fig. 5.13). In southern Oregon during the 2018 fire season, the filters for sleep apnea machines turned black in one night. Even worse than the more local soot from forest fires is the global source of soot that results from diesel motors. Trucks worldwide introduce soot particles, but in Europe and Asia, the dominant use of diesel cars is causing an epidemic problem. European countries have recently recognized the error of promoting the use of diesel cars and are now taking steps to eliminate their use in the future.

The global warming droughts, that are increasing the severity of forest fires, also result in increased pollution from wind-blown dust. I have previously mentioned the dust storms from the Sahara that turn the sky brown and blanket large areas of southern Europe. In March 2018 one of the largest dust storms ever covered Greece and extended to the ski slopes of Russia, which turned the ski slopes reddish brown. These storms not only cause the problem of cleanup, but most important humans are breathing these fine dust particles, which can contribute to lung disease.

6.5   Water Pollution and Limited Supply

The world is already facing a water crisis and nearly half of the human population will face water stress by 2025 (Figs. 5.4b and 6.1 Water use) (International Water Association- www.iwa-network.org). By 2030 the global demand for drinkable fresh water, also called potable water, will exceed supply by 40%. The global water crisis has been made worse because of population growth, increased living standards, global warming and urbanization (Figs. 5.1, 5.4b, I and 6.1 Population). Besides these problems, a fundamental cause of water stress is that the carrying capacity and limiting factor of the Earth for water is worse than any other life-sustaining factor (Fig. 5.4b).Freshwater is only 2.5% of the earth's water and because of glaciers only 0.8% is available as surface or ground water (https://water.usgs. gov/edu/watercycle.html).

India provides an example of the water crisis problems that now are present and will become most severe in Africa and Asia in the future (Figs. 6.4 and 6.5). India already faces severe water stress and carrying capacity because of its limited supply of the global water availability (4%) while supporting 17% of the world's population (https://economictimes.indiatimes.com/news/politics-and-nation/the-precarious-situationof). One billion people there live

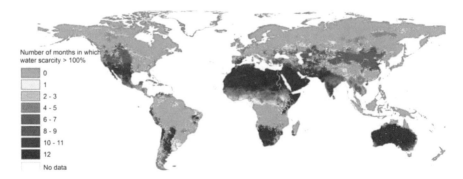

Fig. 6.4 World map showing the number of months in which water scarcity is greater than 100%. Note that more than half of the world's population is affected in western United States, Central America, northern Africa, the Middle East and Australia. Figure   source is https://www.theguardian.com/environment/2016/feb/12/four-billionpeople-face-severe-water-scarcity-new-research-finds

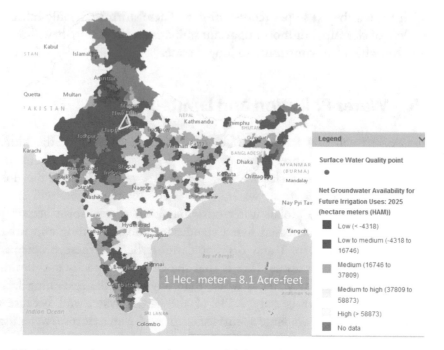

Fig. 6.5 Map showing net groundwater availability in hectare meters (2.47 acres per 1 hectare) for irrigation in 2025 in India. Note that the water scarcity is greatest in northern India where the largest population is, particularly in the golden triangle area near New Delhi that has the densest population. Figure source is http://www.indiawatertool.in/

in water-scarce areas. India also provides an example of the rapidly depleting groundwater supplies that are crucial for drinking water and irrigation (Fig. 6.5). India's rate of groundwater depletion has increased by 23% between 2000 and 2010 (WaterAid). India uses more groundwater than anywhere in the world (24% of the global total). For example the average depth to the ground water table has dropped 18.6 m (62 feet) from 1995 to 2010 in the highly populated Jaipur region of northwestern India (https://www.patrika.com/tags/water-crisis-in-jaipur).

An even more urgent problem is the groundwater pollution by arsenic, nitrates, salinity and fluoride, which has affected tens of millions in India and Bangladesh (The Economist, March 2 to 8, 2019). This contamination becomes greater in deeper groundwater as pollutants seep downward. Fortunately there are solutions to India's water problems such as storing more rainfall (only 6% currently), more efficient irrigation (average efficiently only 31%) and utilizing drip irrigation systems (https://www.waterindia.com/). For example in the Rajesthan region of India, the height of the water table rose 1.38 m (4.66 feet) when 95,192 water conservation structures were built and 2.8 million trees were planted (https://www.hindustantimes.com).

An immediate environmental task for the Earth as part of the UN SDG 2030 goal is to halt the loss of potable water. Lack of clean water is the greatest threat to a sustainable human population because a person will die of thirst or polluted water before hunger (Figs. 5.3b and 6.1 Water use). At present over one billion humans lack access to water and I have observed small African children walking for miles with large canisters of water on their heads to obtain fresh water. In addition, millions of children are dying each year because of contaminated water. Both pollution control and conservation measures need to be implemented to protect our water supplies. These water supplies come from ground water and from surface waters, which can be polluted from the air or pollutants dumped into waterways. Pollution in the air can travel globally and enter surface waters such as (a) with lead from leaded gasoline that spread over continents and across oceans, (b) hydrogen sulfide and mercury from coal burning that have polluted North American, European and Asian lakes, and (c) radioactive pollution from nuclear testing and nuclear power plant accidents that has spread radiation globally and ended up in humans and animals such as Arctic reindeer.

Chemical and mining pollution also can spread for great distances through waterways. Heavy metal pollution from mining and metal refining has also contaminated local areas. For example, near a smelting plant close to San Francisco, California, horses and cows were dying and it was found that particulate matter with toxic heavy metals (e.g. arsenic, lead) from the atmosphere was landing on the grass that the animals were eating. All of the atmospheric and water pollution that destroys potable water can be controlled and eventually eliminated by regulation. There has been good progress at reducing this pollution in North America and Europe. However, the political will must be maintained to keep these regulations and from 2017 to 2020 in the United States there was a move to relax EPA regulations on pollution, even though the majority of the citizens do not want this to happen.

Pollution, overuse and lack of conservation of groundwater are also depleting clean water supplies. At present rates of use, in parts of India, Spain and Italy groundwater could be depleted between 2040 and 2060, and by 2050, as many as 1.8 billion people could live in areas where groundwater levels are fully or nearly depleted (e.g. Fig. 6.5) (https://phys.org/news/2016-12-groundwater-resources-world-depleted-2050s.html). This results from the extensive pumping of groundwater to provide water supplies to populations and mainly for irrigation. Some measures have been taken to flood surface areas with water to slow the depletion of groundwater, but this will not add back the groundwater that has been lost. The California Sustainable Groundwater Act of 2014 is an example of a positive new solution to conserve ground water and can serve as a model for the rest of the world. This act requires California residents to form groundwater sustainability agencies to create their own plans for preserving groundwater. I have talked to a major grower in the California central valley who is actively working with the agency for this area. The plan is working with the farmers making sacrifices and taking responsibility for conserving groundwater.

Preventing pollution of ground water by regulation is another way to preserve potable water. There have been wells drilled to pump pollutants into the ground and this enters the groundwater and makes it unfit for use. Fracking to produce petroleum is an example of this injection of pollution. However, as mentioned earlier, it is not necessary to condemn this process completely as has been done in some states or countries. Instead careful environmental assessment studies can be made to determine where fracking methods will not affect ground water.

Previously I mentioned how increased use of fertilizers spread on land can enter into drainage systems and cause dead areas in the ocean (Figs. 5.4d and 6.1 Fertilizer consumption, 5.2 Nitrogen to coastal zone). The widespread use of fertilizers in the Mississippi River drainage area of the Midwest USA also has contaminated groundwater [4]. Across the northern Midwest there is a groundwater aquifer that extends from the Rocky Mountains to the Mississippi River. In the 1960s, it was noted that in the populated area of the Twin Cities of Minneapolis and St. Paul Minnesota, there were an unusually high number of birth defects and aborted pregnancies. It was found that the groundwater, which supplies much of the drinking water for this area, was contaminated with the nitrates from the use of land fertilizers and the nitrate pollution was linked to the birth defects.

In 1998, I heard a seminar by a U.S. Geological Survey (USGS) scientist that was studying contamination of groundwater under Carson City Nevada. He found that that there were many kinds of prescription drugs in the groundwater supply and people were drinking water that contained a wide variety of drugs such as birth control, antibiotic, painkilling, and animal growth hormone chemicals. At the time I wondered if an area with one of the lowest population densities in the world had this kind pollution, what happened in much more densely populated areas around the rest of the world. Soon after 1998, a number of studies in Europe showed that many surface river and ground waters were highly polluted with prescription drugs. Eventually the USGS and others, in a wide-ranging study throughout the United States, discovered the same thing. The majority of USA surface and groundwaters were contaminated with many prescription drugs that had passed through human bodies (e.g. Kolpin et al. [5]).

There are many harmful side effects for humans and animals because of water pollution with drugs, such as on reproduction and resistance to antibiotics. A 2019 report by the University of York in the UK found that the world's rivers are widely contaminated with antibiotics (e.g. Wilkinson et al. [6]). Researchers analyzed samples from rivers in 72 countries and found that antibiotics were present in 65% of them. This widespread presence of antibiotics and incorporation into human bodies contributes to antimicrobial resistance, which is a global crisis that threatens a century of progress in health [7]. Alarming levels of resistance have been reported in countries of all income levels, with the result that common diseases are becoming untreatable, and lifesaving medical procedures riskier to perform.

A new study suggests that antibiotics not only pollute water and affect human health, but also can alter microbial activity as they enter the soil through animal manure that contains antibiotics given to livestock (Fig. 5.4i) [8]. In the United States alone, livestock already receive an estimated 13 million kg (29 million lbs) of antibiotics every year, which is 80% of the total USA use. The study found that soil microbes consumed carbon less efficiently and released more carbon dioxide into the atmosphere when stressed by certain antibiotics. These microbial effects have implications when assessing long-term soil fertility and greenhouse gas emissions in agricultural fields. In addition genes associated with antibiotics are able to travel widely through airborne particles, migrating birds, storm runoff, and other pathways. Consequently, these environmental impacts may extend far beyond farm fields. This is a significant concern because globally, livestock antibiotic use is projected to increase 67% by the year 2030 (Fig. 5.4i).

Two important ways to decrease polluted water and soil, such as with antibiotics, are to assist countries with techniques for providing uncontaminated water supplies and to develop conservation methods. Unfortunately the California Water Plan provides an example for the lack of conservation planning for water supplies. The rivers of Northern California are dammed and then the water is sent in the huge California aqueduct to the Central Valley and Southern California. The biggest use of this water is for irrigation of crops and two of the main crops are rice and almonds (80% of the world's supply), which require large amounts of irrigation water (The Economist, March 2 to 8, 2019). A great deal of water could be conserved if this water supply were not used to grow water intensive rice or almonds in what is essentially a desert area of the Central Valley of California. This poor choice of growing irrigated, water intensive crops in water stressed areas is global because avocadoes are grown in dry areas of Chile, sugar cane in Pakistan, and rice in India (world's leading exporter) where there is severe water scarcity (Figs. 6.4 and 6.5).

Another major use of the California aqueduct water is for the city of Los Angeles. However this water supply is used once and then dumped into the ocean, whereas smaller amounts of water could be used and recycled as is done with much of the river water supplies in the United States. For example, by the time the Mississippi River water reaches the Gulf of Mexico, it has traveled through the bodies of many human beings https://19january 2017snapshot.epa.gov/www3/region9/water/recycling/. This is a better use of recycling water compared to the one-time use of California aqueduct water. There are examples of tertiary or three stage sewage treatment plants where sewage water can end up as drinking water https://blogs.ei.columbia.edu/2011/04/04/from-wastewater-to-drinking-water/. There's a major potential throughout the globe to apply advanced sewage treatment plants to conserve water.

There are number of both public and private projects to assist developing countries in Africa and Asia to provide potable water or safe drinking water. The first step is to prevent pollution of water supply in these countries and the second step is to increase clean water supplies, such as with simple pumps that will provide well water. One of the major steps to prevent pollution of water is to have sanitation systems that prevent sewage from contaminating drinking water supplies. The worst example of this is India where 600 million people practice open defecation, and the Yamuna River near New Delhi has three million times the recommended safe amount of coliform bacteria (Fig. 6.5) (The Economist, March 2 to 8, 2019). The development of sanitation systems has the additional advantage that it would help eliminate many diseases in these developing countries. There also are water treatment techniques, such as chlorination, that can be used to clean up polluted water supplies and create more potable water.

For undeveloped countries, there are many other beneficial lessons, like water sanitation systems, to be learned from developed countries. The use of drip irrigation can conserve great amounts of water plus prevent soil contamination from conventional irrigation by water flooding. In places like the central valley of California, irrigation has resulted in the destruction of soil from the buildup of salts. The drainage of water from these contaminated soils into some reservoirs has killed nearly all wildlife because of poisonous selenium salts. This potential contamination from selenium was warned about by a USGS colleague of mine years before the California Water Plan for irrigation. Undeveloped countries can learn from other countries mistakes and utilize environmental assessments. They can avoid problems of water and soil pollution from water runoff of stockyards and fields that utilized herbicides, pesticides and too many fertilizers (Fig. 6.1 Fertilizer consumption). Undeveloped countries can reduce the paving over of urban areas that increase storm water runoff with pollutants, flooding, and erosion of soil.

In summary, humans already face a crisis for potable water, however there are solutions to solve this problem for the future (Figs. 5.4b and 6.1 Water use). The United Nations latest annual world water development report notes that more than 25% of humans already live in areas where water is severely scarce and this number will double by 2050 in both undeveloped and developed countries, when taking seasonal variation into account (Fig. 6.4) (The Economist, March 2 to 8, 2019). The volume of water used is already near the maximum that can be sustained without supplies shrinking dangerously. For example a third of the world's biggest groundwater systems, which provide the largest supply of potable water, are in danger of drying out (e.g. Fig. 6.5) (https://phys.org/news/2016-12-groundwater-resources-world-depleted-2050s.html). However, this crisis is solvable because the main cause is water mismanagement, but this will require a huge effort of political will to implement. Although India is the worst example for water management, Israel is an example for the future where water conservation is achieved by utilizing drip irrigation, collecting rainwater, recycling of wastewater and eventually desalinization of ocean water.

6.6   Soil Pollution and Farmland Loss

According to the United Nations IPBES report, 75% of the global terrestrial environment has been severely altered by human actions since pre-industrial times (Fig. 6.2 Domesticated land) and the largest global impact on biological systems has been on land (Fig. 6.2 Terrestrial biodegradation, Tropical forest loss). These land use changes have increased dramatically in just the past half-century. There has been a 300% increase in global food crop production since 1970 and 23% of land areas have reduced agricultural productivity due to land degradation (Fig. 6.2 Terrestrial biodegradation). About 25% of greenhouse gas emissions have been caused by land clearing, crop production and fertilizer use (Fig. 6.1 Fertilizer consumption). Urban areas have grown more than 100% since 1992 and 25 million km of new paved roads are expected by 2050 (Fig. 6.1 Urban population). The amount of renewable and nonrenewable resources that are extracted globally each year has doubled since 1980

(Fig. 6.1 Primary energy use, Water use; Fig. 6.2 Marine fish capture). This includes the unstainable amount of raw timber being harvested, which has increased by almost half since 1970 (Figs. 5.4e and 6.2 Tropical forest loss).

The aforementioned global deforestation is an important contributor to the loss of soil for growing crops. Once land is deforested, there is rapid soil erosion because of the lack of tree roots to hold the soil and keep it from eroding. There are number of conservation methods to keep soil from eroding such as planting along the contours of a hillside rather than up-and-down hillsides, which causes more erosion of the soil. Also the rotation, or changing to different types of crops each year helps to prevent erosion and has the bonus effect of putting nutrients back in the soil. For example in the colonial days in the southern areas of the United States, cotton was planted year after year until the soil was destroyed and then erosion increased because of the lack of planting of new crops.

Now global warming is contributing to the loss of soil because of droughts. The severe droughts that are happening around the globe have resulted in increasing formation of deserts. As more farmland is turned into desert, the wind then erodes the valuable soil that is no longer held by plant roots. I have mentioned previously how the increasing deserts of Africa have caused muddy rains throughout southern Europe. The Gobi desert now is encroaching to within 40 miles of Beijing, China. Saving the soil for farming requires a combination of global efforts to slow and prevent the increase of global warming and deforestation, as well as to institute conservation measures for water.

Chinas soil pollution provides an example of why the world's arable land or farmland is being lost, with dire predictions that it could be gone in 60 years (https://www.scientificamerican.com/article/only-60-years-of-farming-left-if-soil-degradation-continues/). China already is notorious for its air and water pollution as shown by its concentration of air pollutants 10 times greater than WHO maximum recommended levels. Former Prime Minister Wen Jiabao stated that water problems also threaten the survival of the Chinese nation. The Economist (June 10 to 16, 2017) special report provides a detailed description of Chinas problems of loss of arable soil from contamination. A graph shows that there has been a 15% loss of arable land within the last 30 years because of soil erosion and urbanization taking land out of cultivation. In addition, 20% of Chinas farmland has more than the permitted amount of pollutants. At present, China is trying to feed nearly 20% of the world's population on 7% of the global farmland. In the future, at present rates of loss of arable land and soil pollution, within half a century

China will have only about half of their present farmland to feed 1.3 billion people.

The scope of the soil pollution in China is enormous. For example, 250,000 km^2 of polluted land near cities are what is called brown-fields, which is equal in area to the total farmland of Mexico (The Economist, June 10 to 16, 2017). There are also a number of areas set aside for special industrial parks and these have significant problems with chemical contamination. Greenpeace sampled the soil in one of the industrial parks in 2008 and found 226 chemical compounds of which 16 were carcinogenic and three were completely illegal. In 2016 a new university campus was built in one of these industrial park areas that supposedly had been cleaned up and covered with a clay layer. Immediately university students became sick and it was found that there was 80 times the recommended amount of chloral benzene solvent in the soil and groundwater under the university. The university brought a lawsuit against the Chinese government about the contamination and it was thrown out of court.

The Economist (June 10 to 16, 2017) also reported that of the 20% of farmland that has greater than the permitted amount of pollutants, 35,000 km^2 of this farmland is so polluted that it is not used for agriculture. Sixteen percent of the polluted farmland is contaminated with toxic heavy metals such as lead, cadmium and arsenic. Many of these toxic elements are introduced by smelting of ores, chemical plants and irrigation with sewage. Some of this pollution comes from accidents of chemical plants. For example, in the first eight months of 2016, there were 232 accidents such as leaks, spills and fires at chemical plants in China. Other pollutants come from the air pollution of coal-fired power plants, which introduce toxic heavy metals such as mercury, arsenic, cadmium, and lead.

A great deal of Chinas soil pollution, however, results from irrigating land with sewage and industrial wastewater. This irrigation introduces heavy metal pollution as well as organic chemical pollution. In 39 of 55 areas where the polluted soil has been tested, cadmium and arsenic have been found in soil irrigated with sewage water. Of the 60 billion tons of sewage created in China each year, only 10% is treated and thus the majority of this wastewater irrigation is with the polluted raw sewage and industrial wastewater (The Economist, June 10 to 16, 2017).

Pesticides and fertilizer also contaminate the soil, and in 16 provinces of China, 65 pesticides have been found in food in amounts greater than two times the world average (The Economist, June 10 to 16, 2017). Because of the irrigation by sewage water and the runoff from farmland with water containing pesticides and fertilizers, 18% of China's River water is too polluted to use for agriculture, however it is used for irrigation anyway. The reason the polluted river and sewage water is used for irrigation is because in places like North China, there is less water per person than there is in Saudi Arabia. Consequently, the limiting factor of water resources results in irrigation with polluted water.

Why is there this problem of soil pollution and loss of arable land in China and what is the result. Like many of the Communist countries, China had no pollution regulations for decades, there were few scientific studies of pollution, and results of studies were kept secret from the Chinese public. All these aforementioned factors are now being attacked by the Chinese government, because the public has now become aware of the pollution problems of air, water and soil, Unfortunately because of the previous lack of focus on these pollution problems, at present 28% of the rice nationwide has an excess lead content, 10% has excess cadmium, but this reaches 50% in the rice of Guangzhou or the city formerly called Canton (The Economist, June 10 to 16, 2017).

The province of Hunan raises 50% of China's rice and has the area with the most cadmium pollution. The problems of lead contamination that results in mental retardation of children are well known. However, cadmium pollution in rice can result in kidney failure, lung diseases and bone damage. The net result of all the air, water and soil pollution in China is that there are 400–500 cluster village and city areas with higher than average mortality (The Economist, June 10 to 16, 2017). As a result, China in the future may become like the European communist countries where longevity of the population dropped as much as 20 years (https://www.rferl.org/a/life-expectancy-cis-report/24946030.html).

The solution to the soil pollution, loss of farmland and reduced productivity of contaminated farmland in China ranges from extremely formidable to impossible. Soil toxins last for centuries and in one case where farmland was polluted by a chemical spill, 30 years later the soil contains 24 times the permitted level of contaminants. As described in the above and following information from The Economist (June 10 to 16, 2017), the method of covering contaminated soil with an impermeable clay layer or concrete will not suffice as demonstrated by the university constructed on such a re-claimed site with soil contamination. China says it has a plan to make 90% of the polluted soil usable by 2020, but this is not at all realistic. For example, in the famous case of the Love Canal pollution in USA, where 1.2 square kilometers were polluted, it took 21 years to clean up the site. At one contaminated soil site in London, it cost $4000 per square meter to clean up contaminated soil to a depth of one meter. To do a similar type of cleanup for the 250,000 km^2 of contaminated soil in China, the cost would require the total wealth of the entire world. Consequently, it is obvious that this proposed solution in China to clean up 90% of their contaminated soil by 2020 is not possible.

The lessons from China are important to learn for the United States and other countries, because it is obvious that cleaning up widespread soil contamination is impossible. The only way to solve the soil contamination problem is prevent it in the first place. Unfortunately, the United States administration from 2017 to 2021 embarked on a path to cut over 100 EPA pollution regulations for air, water and soil (https://www.cnbc.com/2020/09/03/epa-chief-trump-will-roll-back-more-environmental-rules-if-re-elected.html). For example, the populist USA government eliminated the rules against dumping coal-mining wastes into rivers, and changed the regulations to monitor coal ash ponds containing high quantities of toxic heavy metals.

These changes in regulations potentially can result in millions of tons of pollutants contaminating soil and waterways for thousands of kilometers along their paths. Under the populist USA administration, the Interior Department, conducted a review of its grants, and canceled a $100,000 National Academies of Sciences, Engineering and Medicine study aimed at evaluating the impact of surface mining on nearby communities. If the United States wants to end up like China, it can continue eliminating the EPA and Department of Interior regulations, budgets, scientific grants and scientists that do not agree with their agenda of climate change denial and environmental deregulation.

As examplified by China's soil pollution and USA's reduced pollution regulations, the loss of soil worldwide may be an even greater limiting factor for humans than water supply. Soil erosion and degradation is second only to population growth as the biggest environmental problem the world faces (Fig. 5.4i) [9]. In fact, some experts fear that the world will run out of usable topsoil to grow food within 60 years (https://world.time.com/contributor/ceri-radford-world-economic-forum/). Half of the topsoil on the planet has been lost in the last 150 years and as a result of erosion over the past 40 years, 30 percent of the world's farmland has become unproductive (https://www.worldwild life.org/threats/soil-erosion-and-degradation) [9]. The United States is losing soil at a rate 10 times faster than the soil replenishment rate while China and India, with the worlds largest populations, are losing soil 30 to 40 times faster [9]. Over all, 70% of the earth's topsoil has been degraded [10].

Other aspects of agricultural practices also cause soil degradation. These impacts include deforestation, tilling, fertilization, soil salinity from irrigation, pesticide use, and loss of nutrients, soil microbes and organic matter absorbing CO_2, (Figs. 5.4e and 6.1 Fertilizer consumption, 5.2 Tropical forest loss) (https://www.worldwildlife.org/threats/soil-erosion-and-deg radation). The effects of soil erosion and degradation go beyond the loss of fertile cropland. These effects have led to increased pollution and sedimentation in streams and rivers, clogged these waterways and caused declines in fish and other species. And degraded lands are also often less able to hold onto water, which can worsen flooding. An excellent summary showing the severity of topsoil loss is provided by the Netflix (2020) documentary Kiss the Ground (https://www.netflix.com/es-en/title/81321999).

6.7   Radioactive Pollution of Air, Water and Soil

Radioactive particles are an example of a human-caused global contamination that affects all aspects of the air, water and soil environments. The first large-scale example of this contamination was the use of the atomic bomb on Hiroshima and Nagasaki Japan in 1945. After the immediate deaths of approximately 100,000 people from the explosion and thousands of other deaths from immediate radioactive poisoning in each city, it was realized that there was also radioactive pollution of the air, water and soil. The result of this contamination of all aspects of the environment led to many more thousands of deaths from cancer during the following decades.

In the years immediately following the atomic bomb attacks in Japan, the United States, Russia and France undertook atmospheric H-bomb testing in the 1950s. It soon was determined that radionuclides introduced into the atmosphere by testing were spread around the globe and accumulating in all living things on Earth. For example, reindeer in the Arctic were contaminated by eating lichens with radioactive particles that fell from the atmosphere. Also the radioactive atmospheric contamination changed the entire type of carbon in the atmosphere, which has affected radiocarbon dating since this time. Global anxiety was another result of this testing and the development of Cold War nuclear weapons. As a person growing up during the 1950s, the threat of a nuclear war dominated the daily thoughts of everyone. There was training in all school classrooms for an atomic/hydrogen bomb attack, families built underground bomb shelters in their yards, stored food and obtained gas masks in the event of global atmospheric radioactive contamination from an attack.

Another encounter with radioactive pollution was as a graduate student at Oregon State University in the 1960s. The chemical oceanographers at Oregon State began a large program to study contamination from the Hanford, Washington atomic weapons manufacturing plant that was located several hundred kilometers upstream from the Columbia River mouth [11]. Radioactive waste was polluting the Columbia River, and the waste then drained into the Pacific Ocean off Oregon. The Oregon State scientists found that radioactive particles were absorbed by the microscopic organisms living in the surface waters of the Pacific Ocean. Some of these larval forms of life eventually sunk to the deep ocean floor and carried this contamination to the abyssal depths of the Pacific Ocean off northwest North America. This abyssal ocean contamination as well as the finding of radioactive pollution in Arctic reindeer illustrates how the radioactive contamination has been spread to all global environments.

While I was a graduate student, another example of the severe danger from radioactive pollution was pointed out by a visiting lecturer. Although I do not recall his name, he explained there were 177 buried tanks of high-level radioactive wastes from the USA atomic and H-bomb manufacturing at Hanford, Washington. He also remarked that if the top was lifted off one of these buried storage tanks, the entire Pacific Northwest population would die from radioactive poisoning. So far this has not happened, and probably is not likely to happen, because they are vitrifying or storing these dangerous wastes in glass. Unfortunately, during this cleanup process, just leaking vapors from the tanks has severely sickened many workers cleaning up the contamination at Hanford (https://www.hanfordchallenge.org/chemical-vapor-exp osures). Because the Hanford site contains the world's greatest amount of high-level nuclear waste, it is the most expensive superfund cleanup site in USA, and the ongoing cleanup is estimated to cost 323–677 billion dollars [11]. This example again points out the enormous costs of cleaning up soil pollution and the dangers of not doing this because 1 in 3 USA citizens live near high-level nuclear waste storage areas.

In the 1980s, we had USGS marine geologists studying the Pacific Ocean floor off San Francisco California, because of the large population near nuclear waste. Nearly 50,000 metal drums of 50 gallons each (200 L) with low-level nuclear waste were dumped off California after World War II. Bottom photography showed that 25% of the steel drums had been ruptured and the nuclear wastes were leaking. Because of the continued lack of funding for federal science agencies, to my knowledge, there were never follow-up studies to determine if radioactivity from the leaking drums was spreading to plants and food (e.g. fish, crab) that we humans might be eating from the ocean in this area.

Another similar example of federal budget cuts occurred in 1995 when one of my USGS colleagues was required to stop his study of the radioactive contamination in the Arctic Ocean off Alaska. This contamination resulted from the Russian dumping of nuclear wastes and scuttling of nuclear ships in the Arctic Ocean off Russia. My colleague was an expert in the study of how Arctic ice freezes into the seafloor sediment, which is then picked up in the ice bottom and is carried and re-deposited as the floating ice moves. Because the Arctic ice circulates in a large clockwise pattern, the nuclear wastes off Russia eventually reach the Arctic Ocean seafloor off Alaska. Unfortunately, this contamination has never been studied to understand effects on human and other life, because of USGS budget cuts.

Also during my lifetime, there have been several nuclear power plant accidents, adding to the global effects of radioactive contamination. The first of these was the three-mile-island accident in 1979 in the United States (https://en.wikipedia.org/wiki/Three_Mile_Island_accident). Fortunately the contamination was contained before resulting in widespread pollution. The next major accident, the Chernobyl melt down in Russia in 1986, was much more serious and did cause global contamination of the atmosphere (https://en.wikipedia.org/wiki/Chernobyl_disaster). This accident also caused catastrophic contamination in the local area, where several hundred people immediately died from radioactive poisoning and hundreds more have continued to die from cancer caused by the radioactive pollution decades ago. A large area surrounding this nuclear power plant had to be abandoned for human habitation.

The most recent power plant accident happened with the Fukushima reactors in Japan as a result of the 2011 earthquake and tsunami that destroyed part of the plant (https://en.wikipedia.org/wiki/Fukushima_Daiichi_nuclear_disaster). This was entirely preventable, because there was evidence that tsunamis had previously struck this part of the coast at the nuclear power plant site [12]. Unfortunately, as happens all too often, the evidence of geologists was either ignored or not incorporated into the planning for the nuclear power plant site. I have a power-point slide from a Japanese scientists talk in 2007 that warned of a potential Mw 9 earthquake offshore from the Fukushima power plant.

After the 2011 earthquake, my colleague Dr. Chris Goldfinger, together with Japanese scientists, conducted a more thorough study of previous tsunamis in this Fukushima area. They found that about every 800 years, even larger tsunamis than 2011 inundated the coast up to several kilometers inland from the Fukushima power plant site. From his earthquake frequency studies in Japan and ours off Cascadia, Goldfinger et al. [12] developed the theory that active earthquake faults could have less frequent superquakes then those that typically were expected. In the area of the 2011 Japanese earthquake, previous studies predicted that Mw 8 earthquakes were the largest expected. However,Goldfinger's new studies suggest that Mw 9 earthquakes occur about every 800 years in this area, and any nuclear power plant planning should be prepared for this.

This combination of geological hazards studies and selecting sites for nuclear power plants is an important example of how science can be used to prevent disasters and help a sustainable planet. Geologists have studied the extensive networks of faults that are found throughout California. Thus for many decades, hazards studies have been conducted for nuclear power plant sites. In fact when new active faults have been discovered at a power plant site, some atomic power plants have been shut down in California. This is an excellent example the use of science to prevent nuclear power plant disasters, significant loss of life, and global radioactive pollution.

The Fukushima nuclear power plant disaster and geological hazards assessments point out one of the most outstanding problems to solve for radioactive contamination. Other than prevention of a catastrophic nuclear war or power plant explosion, the most important problem for radioactive contamination is finding permanent storage facilities for high-level nuclear wastes. This problem was highlighted by the Fukushima disaster because the high level waste of the spent fuel rods could not be cooled in the storage ponds next to the power plant. At all nuclear power plants of the world, the contaminated rods are stored in ponds by the power plants because there is no permanent storage for them. Storage requires keeping rods out of the environment for hundreds of thousands of years. At present, nuclear power plants become unusable after a few decades because of the high level radioactive wastes that are stored on site rather than in permanent storage. The USA spent hundreds of millions of dollars and decades of geological research to select a permanent storage site in Nevada. Unfortunately, the citizens of Nevada refused to permit using it, so the problem remains unsolved for the USA and globally, waiting for the next nuclear power plant disaster.

A problem of radioactive contamination that has mainly been solved is the use of X-ray equipment and other radioactive materials. Use of radium to make watch dials glow, was one of the first examples of this contamination. In the early twentieth century when many workers in watch factories began dying from radioactive poisoning, this practice was stopped. In the 1950s, we used to go to shoe stores and use X-rays to look at how our shoes fit. Later it was found that such use sometimes was linked to thyroid malfunction and cancers, from which some of my friends suffered. The biggest problem was from use of unshielded X-ray machines for many medical, dental, industrial and scientific purposes. Some of my geologist colleagues from their mid-twentieth century research contracted cancer and died because of the use of poorly protected X-ray machines. However, by the time of my research with such X-ray machines, safety measures worldwide had been introduced. Machines and users were shielded with lead to absorb X-rays, workers wore badges to measure their exposure to X-ray radiation, and laboratories with X-ray machines had numerous warning signs posted.

6.8   Ocean Pollution, Fisheries and Coastal Habitat Loss

According to the United Nations IPBES 2019 report [15], 40% of global marine environments have been severely altered by human actions since pre-industrial times and the second largest global impact on biological systems has been in the ocean. Pollution entering coastal ecosystems has produced more than 400 ocean "dead zones," totaling an area bigger than the United Kingdom [15] (Figs. 5.4d and 6.2 Nitrogen to coastal zone). These areas are so starved of oxygen they can barely support marine life. Marine plastic pollution alone has increased ten times since 1980 with an average of 300–400 million tons of waste dumped into the world's waters annually. Another human impact is that unsustainable industrial fishing results in 55% of the total ocean area (Figs. 5.4c and 6.2 Marine fish capture).

David Balton, former deputy assistant secretary for oceans and fisheries in the U.S. Department of State testified before a US senate hearing in 2018. He said Illegal, Unregulated, and Unreported (IUU) fishing remains a serious threat to fisheries worldwide because of the interrelated challenges of unsustainable fisheries, marine pollution, global warming, ocean acidification (see Chapter 4 Sect. 4.3), and other phenomena (Figs. 5.2, 5.4c, d and 6.2 Ocean acidification, Nitrogen to coastal zone). He added that it's not only for the USA to protect its own fisheries, because migratory species coupled with the aforementioned changing ocean conditions, means that we have to embrace the global nature of the challenge facing this critical resource for human food.

IUU fishing may total more than $36.4 billion annually in black market profits and accounts for one out of every five fish caught globally, according to the Stimson Center in Washington, D. C. For example, When the U.S. and Chinese coast guards in June 2018 jointly intercepted a Chinese fishing vessel in the North Pacific Ocean, the coast guard crews discovered almost 73 metric tons of illegally caught salmon. That cooperative effort gave a glimpse of the IUU fishing, which causes economic and ecological damage to fisheries. However, there is a sign that countries can work together globally on the issue and in 2016 the Port State Measures Agreement (PSMA) was put in force by 55 countries to combat IUU fishing.

Similar to other biological global changes, the route to sustainable wild fisheries again is best achieved by global conservation measures and prevention of pollution plus global warming. The National Marine Fisheries of United States has developed an excellent program for sustainable fisheries. Fishing stocks and fishing catches are monitored and the amount of fishing that is allowed depends on maintaining a sustainable stock of fish. Another major program for oceans of the United States territories is the development of Marine Protected Areas of the ocean so that marine species can be conserved, nurtured and sustained. These protected areas also have been established offshore from some areas of Europe and need to be put in effect in many other countries of the world. Unfortunately, in 2018 the U.S. president issued an executive order reducing marine protected areas, whereas the USA should increase areas and provide scientists to help implement these programs for other countries (https://www.whitehouse.gov/presidential-act ions/executive-order-regarding-ocean-policy-advance-economic-security-env ironmental-interests-united-states/).

The marine environment, which is 70% of the Earth, has had less study and conservation compared to the continental areas. To implement sustainable fisheries and marine sanctuaries there needs to be much more scientific assessment of the world's oceans. Only a small percentage of the ocean area has been explored or mapped. Study of the ocean needs to be greatly increased to fully understand the fisheries and where marine sanctuaries need to be established to nourish these fisheries. Major fisheries like that for the Grand Banks cod have completely disappeared because not enough studies have been made and sustainable fisheries have not been established globally (Fig. 5.4c). In contrast to these needs, the United States federal government fleet of marine research vessels has been reduced since the 1970s. Consequently, less ocean research has been done by federal scientists since then.

Although the marine area of the Earth dwarfs the continental areas, pollution has affected all of the oceans. Until recently, sewage and trash have been thrown uncontrollably into the ocean. Smaller enclosed seas, like the Mediterranean Sea, eventually became so polluted that it was dangerous to swim along much of the Mediterranean coast. However, a major effort was undertaken and countries around the Mediterranean have cooperated to reduce the pollution.

Unfortunately, many types of ocean pollution have continued to spread throughout the seas from the dumping of trash, particularly plastics. The production of plastics has quadrupled in the past four decades and if this trend continues, by 2050 the manufacture of plastics will create 15% of the world's CO_2 emissions, the same as the global forms of transportation now do (CNN editorial, April 17. 2019). Because of the great increase of plastics, around 150 million metric tons of plastic are already floating in our oceans and an additional eight million tons enter the water each year (www3. weforum.org/docs/WEF_The_New_Plastics_Economy.pdf). In addition, it is estimated that there is 14 million metric tons of microplastics sitting on the ocean floor [15]. Microplastics are pieces of plastic that have been worn down by the elements into tiny fragments, smaller than 5 mm (0.19 inch).

Another result of the increased production of plastics is that there are huge areas in the Atlantic and Pacific oceans where the circulation is sluggish and masses of trash have collected in the open ocean. Because nearly every piece of plastic ever made still exists today, more than five trillion pieces of plastic are already in the oceans, and by 2050 the weight of plastic in the sea will be more than fish, according to the Ellen MacArthur Foundation (https://www.businessinsider.com/plastic-in-ocean-outweighs-fish-evidence-report-2017-1?IR=T). Also, millions of tons of trash have collected on beaches and remained in remote uninhabited islands like Midway and Henderson. For example, Henderson Island, between New Zealand and Chile, is covered in an estimated 38 million pieces of trash (https://www.nationalgeographic.com/news/2017/05/henderson-island-pitcairn-trash-plastic-pollution/).

Lebreton et al. [14] report that the sprawling Great Pacific Garbage Patch (GPGP) in the eastern North Pacific Ocean not only is the largest plastic accumulation zone on Earth, but also is contaminated with 4–16 times greater amount of floating plastic by mass than previously estimated. The 1.6-million-square-kilometer garbage patch (3 times the area of France) contains at least 79,000 metric tons and an estimated 1.8 trillion pieces of floating ocean plastic. This increased estimate of GPGP plastic resulted from extensive new research by the Ocean Cleanup Foundation involving 30 vessels and a 2016 aerial expedition. On the positive side this Foundation has an ambitious goal of cleaning up 90% of plastic in ocean gyres by 2040.

Introduction of petroleum also causes ocean pollution more locally in coastal areas, such as from the Horizon oil spill in the Gulf of Mexico, and worldwide from the practice of cleaning out bilge tanks of oil tankers in the open sea. Microscopic nodules of petroleum have spread throughout the ocean and there have been few studies to determine the effect on the microscopic plankton, birds and animals.

Recently it has been discovered that microscopic plastic balls or nano-plastics that are introduced into many products such as cosmetic creams and toothpaste are being incorporated into the bodies of birds, fish and mammals throughout the ocean, and also in humans eating fish (e.g. Wilkinson et al. [6]). The effects of this are unknown for humans and other species. However, eliminating the use of nano-plastics in products can solve this problem and many companies are now taking these steps.

Use of fertilizers to grow crops is another widespread problem of pollution in the ocean that has previously been mentioned (Fig. 6.1 Fertilizer consumption, 5.2 Nitrogen to coastal zone). These fertilizers wash off the land into the river drainage systems and these eventually enter the ocean. An example of this is the Mississippi River, which drains rich croplands throughout the middle of North America. The extensive use of fertilizers on these croplands results in ammonium nitrate fertilizer drainage into the Mississippi River, which then drains into the northern Gulf of Mexico. These fertilizers drained down the river then fertilize the ocean water, which results in overproduction of the microscopic plants in the sea. With this large growth of algae, there becomes an overproduction of the animal life, which utilizes all of the oxygen in the seawater. With this lack of oxygen during the summer, there is a huge die off of many fish and other animal species (https://www.latimes.com/local/la-me-deadzone15feb15-story.html). As a result, an area the size of New Jersey in the northern Gulf of Mexico has become a dead sea with the death of most of the animal life. These dead zones also happen globally in 400 other locations of the ocean [15] and the overuse of fertilizers and introduction of other pollutants needs to be reduced to prevent this die off of significant areas of the ocean (Fig. 5.4d).

A similar extensive die off of marine animal species also takes place as a secondary effect of global warming. An example of this occurred in 2002 during our scientific cruise off Oregon and Northern California to study earthquake history. While we were 300 km offshore from the coast of California, in beautiful high-pressure sunny weather, we encountered high winds of nearly 60 miles per hour (100 km per hour) and large waves. This wind was caused by the extremely hot weather in the Central Valley of California, which resulted in heating of the air and convection cells of the rising hot air over the Central Valley. This unusually high amount of rising hot air had to be replaced, which caused the strong onshore winds that affected our ship hundreds of kilometers from the shoreline.

The other most important result was that these onshore winds caused extremely strong currents along the shoreline, which moved the surface water offshore. This near-surface water then had to be replaced by deeper upwelling water, which is rich in nutrients that support the ocean life. Again there was anomalous overproduction of organisms that depleted the water of oxygen and caused dead zones. As a result there was the greatest die off of ocean animal life that has occurred during historical time, and the Pacific Northwest coast beaches of North America were littered with dead fish and other bottom dwelling animals such as crabs (Fig. 6.7) (https://www.latimes.com/local/la-me-deadzone15fcb15-story.html). Of course the solution to these extreme die offs in the ocean is to limit global warming.

The most severe problems for pollution and lack of conservation in the oceans have been in coastal areas. For example, in the 1970's I saw an excellent presentation by Dr. Miles Hayes from University of South Carolina that compared the pristine coastal areas of Alaska with the highly polluted coastline of New Jersey where there are numerous chemical plants. Similar areas like New Jersey exist around the world, particularly near population centers. The USA has instituted a number of regulations to stop these types of pollution, but the administration from 2017 to 2020 has eliminated some regulations against dumping chemical wastes, such as from coal mining into rivers. The opposite should be happening to eliminate pollution in coastal areas. The USA also should help other countries around the world to implement the techniques that we have developed to cut pollution in the United States.

In addition to chemical pollution in coastal areas, there has been significant destruction of coastal biological environments, such as mangroves and marshlands, which are called blue carbon systems. Because these environments are extremely productive biologically, they are important for taking CO_2 out of the atmosphere. Unfortunately besides pollution, huge areas of coastal wetlands and estuary marshlands have been reclaimed for urban development. An example of this is the salt evaporation ponds in San Francisco Bay, the largest western North American estuary. Fortunately, some coastal areas like San Francisco Bay have reduced sewage pollution and are now restoring some salt pond areas back into natural marshlands, although large areas of the Bays marshlands still remain filled in and covered with urban development. The solution to the destruction of coastal environments is the development of marine coastal sanctuaries and preventing urban development in these areas.

A recent study on coastal blue carbon systems shows how important this loss of the earth's mangrove forests and marshlands are for global change problems (https://www.estuaries.org/images/Blue_Carbon). Mangroves are one of the most important parts of the blue carbon system, which also includes sea grasses and salt marshes, all of which are known to be the most biologically productive environments per unit area on Earth. This new study points out how important these systems are for the CO_2 sequestration or uptake of carbon dioxide out of the atmosphere, which helps the problem of global warming caused by human increased production of CO_2. Blue carbon ecosystems remove 10 times more CO_2 per hectare (2.5 acres) from the atmosphere than inland forests. www.estuaries.org/images/Blue_Carbon/rae%20blue%20carbon%20factsheet_30oct2014.pdf.

Scientists have found that the mangroves, marshes and sea-grass incorporate more CO_2 for a longer time compared to other marine microscopic plants or phytoplankton and kelp. The reason they take more CO_2 out of the atmosphere for a longer time is because 50 to 90% of coastal blue carbon storage takes place in the sediment and not in the plants [16]. The plant debris is trapped and buried in the bottom sediment and thus you can dig down several meters in the sediment and still encounter plant leaves and other plant debris of sequestered carbon. In contrast, the coastal kelp forests and phytoplankton do not get incorporated into sediment and rapidly decay to return the CO_2 to the atmosphere. Scientists also found that the mangroves, tidal marshes and sea grasses store up to 35 times more carbon per unit area relative to phytoplankton CO_2 [16].

The loss of the mangroves is critical because as they disappear, soils are washed away into the sea and the carbon is returned to the atmosphere. Also the mangrove forests on the coast are good buffers against the increasingly strong storms caused by global warming [17]. In addition to helping offset carbon emissions and providing a storm buffer, the mangroves are valuable habitat and productive nurseries for many marine species. I mentioned before that in 1966, the mangroves in Malaysia had extended 50 km into the sea since the Marco Polo visit around 1200 A.D. Now the reverse is taking place compared to the mangrove growth after the visit of Marco Polo, and the storm buffer and mangroves are disappearing.

At the present rate of 50% loss in the past 50 years, in another 50 years most of the mangrove forests will have disappeared [17]. This is similar to other ocean resources such as wild fisheries and coral reefs that are estimated to disappear at about the same time, if these global changes continue at the present rate. A global program for a blue carbon initiative has begun to save the vanishing mangroves and other coastal system, to preserve habitat and marshlands, and to help slow the effects of global warming (United Nations Environment Program (UNEP). Unfortunately of the 150 countries that have coastal wetlands and signed the Paris Agreement, only 29 have joined the blue carbon initiative. Obviously to help a sustainable Earth, we need to encourage all countries with coasts to join the blue carbon initiative, and be motivated to save coastal environments.

The destruction and loss of coral reefs in the next half-century is another coastal environment facing multiple global changes. These changes are related to many human activities even though hundreds of millions of humans are dependent on reefs for a food source, tourism income and protection from storms along 150,000 km (93,000 miles) of shoreline for more than 100 countries (The Economist, October 28, 2019). Reefs not only are a spectacular environment to observe and important for tourism, but they also are a critical food resource for 130 million people. As the food resource from reefs has been reduced from over fishing and by destructive global changes, some fishermen have resorted to dynamiting local sections of reefs to stun and kill all the fish.

A much more pervasive destruction of reefs has resulted from coral bleaching caused by rising ocean temperatures which kill the algae that corals feed on (Fig. 5.2). Corals bleach when they experience temperatures above their normal summer maximum for a month or two. The BBC blue planet II documentary reports that 50% of the world's coral reefs are affected by bleaching. In 2016, the Guardian newspaper reported that 93% of the world's largest 2,300 km long Great Barrier Reef off Australia was affected by bleaching as well as 75% of all the worlds reefs. This bleaching was particularly severe in 2016, when an El Niño event combined with global warming to raise ocean temperatures. You can see the destruction of coral reefs yourself by looking through glass bottom boats or snorkeling on coral reefs.

Although the 2016 El Niño contributed to severe coral bleaching, global warming is the long-term cause (e.g. Hughs et al. [18]). The three worst historical mass bleaching have occurred in 2016, 2017 and 2020, and the bleaching has spread throughout the entire Great Barrier Reef (https://www.coralcoe.org.au/news-and-media/media-releases). The Economist (October 28, 2019) reports that bleaching now is happening 5 times faster than it did in the 1970s. In the two previous 1998 and 2002 events, 40% of the Great Barrier Reef escaped bleaching and only 18% experienced severe bleaching. In 2016, 55% of the Reef experienced severe bleaching and of 911 reefs surveyed only 68 escaped bleaching entirely. If coral remains bleached for an extended period, it is likely to die and the northern Reef already has experienced mortality as high as 50%. In the future, the IPCC [19] estimates that 70–90% of the world's reefs will decline if average global temperatures reach 1.5 °C (2.7 °F) greater than pre-industrial times. The newest report in 2020 indicates that all reefs may disappear by 2100 (https://news.agu.org/press-release/warming-acidic-oceans-may-nearly-eliminate-coral-reef-habitats-by-2100/).

The ocean acidification, because of increased CO_2 dissolving in the ocean, is another global change threatening coral reefs (see Chap. 4 Sect. 5.3). As the ocean becomes more acid, the ability to form carbonate skeletons is more difficult for corals. Corals also need protection from other local sources of human effects, such as sewage pollution, farm runoff, and dumping of construction wastes plus plastics into the ocean. In sum, reef destruction is another example of the multiple human-caused global changes that combine to destroy an environment.

6.9   Population Growth

The accelerating growth of global population is the bottom line for all of these human-caused global change problems of energy use, global warming, over fishing, pollution of air, water and soil, plus depletion of the sustainable resources for humans (Figs. 5.4i and 6.1 Population). People have been warned about the population growth since the time of Malthus in the 1700s (Fig. 1.2) [20]. Fortunately, the contributions of science and technology have been able to keep up more or less with the growth of population. However, the pressure on our Earth's resources results in much of the world population existing below the poverty line and living barely sustainable lives.

The estimates that at present rates of use, the ocean fisheries, groundwater, coral reefs, mangroves etc. will be gone in the next half century, and that approximately 50% of the present potable water and arable soil will be depleted, while the human population grows 50%, does not bode well for a sustainable population. At a minimum the world will become much less secure as countries fight over these depleted resources [2], for example, Diamond [26] has already pointed out in his book Collapse, how some of the recent genocides such as in Rwanda have resulted where population density is extremely high and environmental resources are under severe stress. Also, the civil war in Syria has followed a time of severe drought and loss of crops in Syria that has been intensified because of global warming [21]. However, Selbya et al. [22] point out that the many complexities make it difficult to attribute part of the cause for the Syrian war to drought and doing this may just enforce global warming skeptics. Nevertheless, we undoubtedly face the prospect of future conflicts, particularly because of our depleting water resources, droughts and soil loss.

China, with the world's largest population recognized these problems related to population growth, and with an autocratic political system instituted population control with a one-child policy. This was effective but also resulted in abuses such as controlling the sex of children and prohibiting the birth of girls. Now the policy has been changed, however birth rates remain low. Although ideally population control policies like that of China would be effective in controlling population growth, this certainly is not practical considering the many different cultures and religious practices in the world's population, as well as the loss of personal freedom. The most effective means of reducing population growth is to reduce poverty. The developed countries of the world already have birth rates for a sustainable population and most developing countries, except Africa, are approaching this goal now (Fig. 6.6). If Africa also reaches this goal by 2050, the Earth may be able to stabilize at a population less than 10 billion people and match the earths estimated carrying capacity for humans [23].

There are practical measures such as education of the world's population concerning the dire need for reduced population growth, if we want

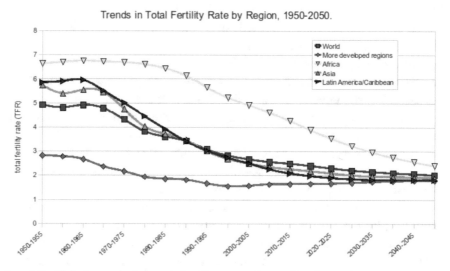

Fig. 6.6 Global trends in total fertility rates by region from the year 1950 and projected until 2050. In 2020 the more developed world has reached sustainable rates and most other regions except Africa are close to sustainable rates where there is no net population growth. However, by 2050 the population is estimated to surpass 9 billion [24], which is the estimated carrying capacity of the earth's resources [23]. Figure source is [25] and https//commons.wikipedia.org/wiki/File:Trends_in_TFR_1950-2050.png

Fig. 6.7 Photograph showing dead dungeness crabs in the tidal zone of the Oregon coast USA. This is an example of the massive die off of marine organisms along the USA Pacific Northwest coast in 2002. This die off was caused by unusual atmospheric and ocean circulation related to global warming (see Chap. 5 Sect. 5.8 for an explanation). Similar die offs are found in hundreds of dead zones in the ocean caused by fertilizer runoff from rivers into the sea (Fig. 6.1 Fertilizer use) [15]. Figure source is https://www.latimes.com/local/la-me-deadzone15feb15-story.html

to sustain human life on our Earth. There can be education for family planning and for birth control measures. This will require significant changes in religious and political philosophies if we want to sustain a quality of human life on our Earth in the future. Preventing birth control measures will increase population growth and the spread of diseases like AIDS. Recent changes in political philosophies, such as after the 2016 election in the United States, have resulted in the reduction or elimination of funding for family planning clinics and dissemination of birth control methods. Such political philosophies are extremely dangerous for the future of our planet and shows that the most important efforts of education are for political and religious leaders to be aware that their policies are threatening the future of human life on our Earth.

The coronovirus pandemic in 2020 provides a huge wakeup call related to population growth. Exponential population growth has resulted in human encroachment into nature, which brings people and livestock into closer contact with wild animals. As Fareed Zakaria also has pointed out in his Global Briefing on May 8 2020, this encroachment and antibiotic-heavy factory farming offers a perfect environment for pathogens to develop. The real scandal, Fareed notes, is not for politicians to blame China for Covid-19, but what we together are doing to the Earth because of population growth, and what only we together can accomplish for the common good of humanity on the Earth. The solution will require political leadership and education for the common good of humans (see discussion in Chap. 5 Sect. 5.10).

6.10   Lake Tanganyika and Combined Global Change

Dr. Andrew Cohen, as quoted here from the Conversation in May 2017 (https://theconversation.com/the-fate-of-africas-lake-tanganyika-lies-in-the-balance-75467), has provided the following excellent example of how multiple human-caused global changes can combine to affect the sustainability of a specific environment. As he points out, Lake Tanganyika is changing, and the fate of millions lies in the balance. This example goes back to my original global change example of how small lake environments are the most vulnerable and first to show effects of global change. It also shows how global changes combined with population pressures can destroy a sustainable environment. It serves as a microscopic example of how our entire Earth faces immense pressures of human caused global change that threaten our entire planet and the sustainability of human life.

Lake Tanganyika occupies 39,000 km2 in the African Rift Valley and borders Tanzania, the Democratic Republic of the Congo, Burundi, and Zambia. It's one of the oldest lakes in the world, probably dating back about 10 million years. That expanse of geological time has permitted hundreds of unique species of fish and invertebrates to evolve in isolation and every day millions of people rely on the lake for food. However, despite being a world-class reservoir of biodiversity, food and economic activity, the lake is changing rapidly and faces an uncertain future. Lake Tanganyika was recently declared the "Threatened Lake of 2017" because it is adversely affected by human activity in the form of deforestation, overfishing, population growth and rising temperature from global warming.

Beginning in the late 1980s scientists studying the lake began to notice significant and concerning changes caused by human activity. Evidence showed that underwater habitat degradation was taking place adjacent to hill slopes where rapid deforestation was taking place for agricultural lands and for urban expansion of the fast growing population centers around the lake (Fig. 5 Terrestrial biosphere degradation). This activity has led to a rapid increase in the amount of loose sand and mud being washed into the lake, which is smothering the lake floor.

The present living biota of Lake Tanganyika can be imagined like a thin bathtub ring that hugs the shallow zones around the deep (up to

1470 m 4778 feet) lake. The hundreds of species that inhabit the sunlit shallow water give way to a dark expanse of water lacking oxygen and animal life. Eroded sediments are being carried into the lake, affecting this strip of biological activity. Scientists have begun to document where the impact is being felt and also are looking back in time by collecting sediment cores with fossils of the many unique animals to see when the impact was first felt. They have found that some heavily populated regions lost much of their diversity more than 150 years ago and other more southern parts of the lake are seeing these effects only in recent decades.

Excess sedimentation is just one problem. Fishing pressure and global warming is also affecting the lake. Large scale fisheries for the lake's small sardines started in the 1950s, and quickly mushroomed into a major industry. Up to 200,000 tons of fish per year are exported and make up a large portion of the average person's animal protein intake in the surrounding regions. In recent years this fishing yield has declined dramatically. This has been partially caused by the unsustainable growth in fisheries, and exacerbated by large numbers of refugees flooding into the region because of conflicts in Rwanda, Burundi and the Democratic Republic of Congo during the 1990s.

It's now clear that global warming is another factor threatening the lake's fragile ecosystems and lake sustainability. Starting in the early 2000s, scientists began to document that the surface waters of Lake Tanganyika, like Lake Tahoe, were warming rapidly [3]. This is most likely because of global climate change related to an increase in greenhouse gas emissions. Warm water is relatively light and in tropical areas mixes poorly with the deeper layers of the lake. This in turn keeps the vast pools of nutrients from being churned back to the surface by waves. The warming water and lack of circulation cuts down on the growth of floating plankton, which many of the lake's fish eat.

Scientists have been able to show that the decline in fish populations began well before the onset of commercial fishery in the 1950s. This implicates climate change and lake warming as the probable cause for much of the fishery's long-term decline. This trend is not likely to be reversed as long as global warming continues. A related consequence of the warming and reduction of mixing in the lake, is a continuous shallower transition from the oxygenated to deoxygenated waters on the lake floor. This means there is an even shallower oxygenated ring for the production of fish and a further reduction of the food resource for the rapidly expanding population.

Another potential new problem has emerged which is exploration for oil and gas. Rift lake sediments of the type found in Lake Tanganyika are well known among geologists as reservoirs of hydrocarbons, because over millions of years, vast quantities of plankton have died and settled on the lake floor.

The deposits of plankton decay in the sediment layers and eventually turn into petroleum. The consequences of any possible petroleum production in Lake Tanganyika are still unknown. But the record of catastrophic oil spills, along the Niger River Delta, highlight the critical need for careful study and environmental planning before any possible petroleum production proceeds in the fragile Lake Tanganyika. The biological and economic riches produced by 10 million years of evolution could lie in the balance.

References

1. Steffen W, Broadgate W, Deutsch L, Gaffncy O, Ludwig C (2015) The trajectory of the Anthropocene: the great acceleration. The Anthropocene Rev 2:81–98
2. Nelson CH, Lamothe PJ (1993) Heavy metal anomalies in the Tinto and Odiel river and estuary system. In: Kuwabara JS, Trace Contaminants and Nutrients in Estuaries. Estuaries 16:496–511
3. Schladow G (2018) Annual lake Tahoe state of the lake report. UC Davis Tahoe Environmental Research Center, 97p
4. Power JF, Schepers JS (1989) Nitrate contamination of groundwater in North America. Agr Ecosyst Environ 26:165–187
5. Kolpin DW, Furlong ET, Meyer MT, Thurman EM, Zaugg SD, Barber LB, Buxtpn HT (2002) Pharmaceuticals, hormones, and other organic wastewater contaminants in U.S. streams, 1999–2000: a national reconnaissance. Environ Sci Technol 36:1202–1211. https://doi.org/10.1021/es011055j
6. Wilkinson J, Hooda PS, Barker J, Barton S, Swinden J (2017) Occurrence, fate and transformation of emerging contaminants in water: an overarching review of the field. Environ Pollut 231:954–970. https://doi.org/10.1016/j.env pol.2017.08.032
7. United Nations Interagency coordination group on antimicrobial resistance (IACG) (2019) No time to wait: securing the future from drug-resistant infections, Report to the secretary-general of the United Nations, 28 p
8. Webking C, Badgley B, Barrett JE, Knowlton KF, Lucas JM, Minick KJ, Ray PP, Shawver SE, Strickland MS (2019) Prolonged exposure to manure from livestock-administered antibiotics decreases ecosystem carbon-use efficiency and alters nitrogen cycling. Ecol Lett 22:2067–2076.https://doi.org/10.1111/ele.13390
9. Pimentel D (2006) Soil erosion: a food and environmental threat . J Environ Dev Sustain 8:119–137

10. Dregne HE (2002) Land degradation in the drylands. Arid Land Res Manage 16:99–132. https://doi.org/10.1080/153249802317304422
11. Hanford (2019) Hanford lifecycle scope, schedule and cost report, DOE/RL-2018, 16 p
12. Goldfinger C, Ikeda Y, Yeats RS, Ren J (2013) Superquakes and supercycles. Seismol Res Lett 84:24–32. https://doi.org/10.1785/0220110135
13. Barrett J, Chase Z, Zhang J, Banaszak Holl MM, Willis K, Willians A, Hardesty BD, Wilcox C (2020) Microplastic pollution in deep-sea sediments from the great Australian bight. Front Mar Sci. https://doi.org/10.3389/fmars.2020.576170
14. Lebreton L, Stat B, Ferrari F, Sainte-Rose B, Aitken J, Marthouse R, Hajbane S, Cunsolo S, Schwarz A, Levivier A, Noble K (2018) Evidence that the Great Pacific Garbage Patch is rapidly accumulating plastic. Sci Rep 84666. https://doi.org/10.1038/s41598-018-22939-w
15. United Nations IPBES (2019) Intergovernmental science-policy platform on biodiversity and ecosystem services (IPBES) Global assessment summary for policymakers. https://www.ipbes.net/news/ipbes-global-assessment-summary-policymakers-pdf
16. Witman S (2017) Coastal wetlands effectively sequester "blue carbon". EOS 98. https://doi.org/10.1029/2017EO079397. Published on 21 August 2017
17. Cleverley M (2015) Amazing mangroves. Pacific Ecologist 23:28–31
18. Hughes T, Kerry J, Baird A, Connolly S, Dietzel A, Eakin C, Heron S, Hoey A, Hoogenboom M, Liu G, McWilliam M, Pears R, Pratchett M, Skirving W, Stella J, Torda G (2018) Global warming transforms coral reef assemblages. Nature 556:492–496
19. IPCC (2019) Special Report on the Ocean and Cryosphere in a Changing Climate. https://www.ipcc.chsrocc
20. Malthus TR (1798) An essay on the principle of population, anonomous essay later identified with Malthus as the author
21. Gleick P (2014) Water, drought, climate change, and conflict in Syria Weather. Climate Soc 6:331–340
22. Selbya J, Dahib OS, Frohlichc C, Hulme M (2017) Climate change and the Syrian civil war revisited. Polit Geogr 60:232–244
23. Wilson EO (2002) The future of life. Knopf Doubleday Publishing, New York City, p 220
24. United Nations (2019) World population prospects, the 2008 revision
25. United Nations (2008) World population prospects 2008
26. Diamond J (2005) Collapse. Viking Press, p 592
27. Heal GM (2016) Endangered economies: how the neglect of nature threatens our prosperity, Columbia University Press, New York, p. 227. https://cup.columbia.edu/book/endangered-economies/9780231180849
28. Sieminski A (2013) International energy outlook, 2013, U.S. Energy Information Administration, Washington, DC, p. 1–33

29. AAPG (2017) Search and Discovery Articles: Tinker #70,268; Koonin #70,269; Yielding #70,270; Snell #70,271; Ausubel #70,272; Medlock III #70,273
30. Netflex (2020a) Kiss the ground. https://www.netflix.com/es-en/title/81321999
31. Netflix (2021) David Attenborough Johan rockström breaking boundaries: The science of our planet. https://www.netflix.com/es-en/title/81336476
32. Netflix (2020b) Brave Blue World. https://www.netflix.com/es-en/title/813 26710

Part III

Our Choice

7

Solutions for a Sustainable Earth

7.1   Status of Global Change Education

The best hope for a sustainable human race on our Earth is education about all aspects of human-caused global change. The global changes besides warming, that I have described, appears to be rarely taught except for advanced students majoring in the environmental sciences in universities and graduate schools.

The most important educational goal for global change is to reach students during their required education in grade and high schools. At present the main aspect of global change that is taught within public education is climate change and even this appears to be seldom in the USA, as reported in a CNN editorial (https://edition.cnn.com/2017/06/14/health/climate-change-schools-partner/index.html). This is improving with new science teachers and in more USA progressive schools and appears to be better in European schools, which require more science education in high schools than the USA. The result of this better science education in Europe is agreement between liberal and conservative governments that global warming is a threat to our Earth.

Unfortunately, the USA populist government from 2017 to 2020 dropped out of the Paris Agreement and the population has a greater amount of climate change denial than Europe. The way to change this opinion is through science education so that the general population can overcome the rhetoric of climate change denial. As a result of denial, disregard for science education has reached down to some state educational systems, where

© The Author(s), under exclusive license to Springer Nature
Switzerland AG 2021
C. H. Nelson, *Witness To A Changing Earth*,
https://doi.org/10.1007/978-3-030-71811-4_7

science in general and specifically denial of climate change is encouraged. For example, some schools teach climate change as if it were an ongoing debate whereas the most recent analysis of peer reviewed scientific articles shows that 100% of climate scientists agree that global warming is caused by humans [1].

There is pending and failed state legislation pushing teachers to be less declarative on climate change and evolution. In 2017, CNN reported that 11 bills designed to alter science education standards in states have been unsuccessfully introduced across the United States. (https://edition.cnn.com/2017/06/14/health/climate-change-schools-partner/index.html). For example, a bill in the Texas House of Representatives would allow science teachers to teach "the scientific strengths and scientific weakneses of existing scientific theories," namely theories around subjects such as climate change, evolution, the origins of life and cloning. The bill maintains, "The protection of a teacher's academic freedom is necessary to enable the teacher to provide effective instruction".

According to CNN, the Texas measure mirrors efforts in Idaho and West Virginia, where objections to the inclusion of climate change in state education standards have met with varying degrees of success (https://edition.cnn.com/2017/06/14/health/climate-change-schools-partner/index.html). There is also a bill in Florida that would make it easier for residents to challenge school textbooks, including those that discuss topics such as climate change and evolution. That class of bill is couched in the language of academic freedom," said David Evans, executive director of the National Science Teachers Association. "But teachers shouldn't be permitted to teach things in class that aren't science." Most of these bills in other states have been defeated. In Iowa, legislators voted down a measure that would have prohibited the state Board of Education from adopting the Next Generation Science Standards, which include climate change in the curriculum.

In addition to the negative effect of climate change denial by some national and local politicians, according to CNN, there appears to be a lack of knowledge and teaching of the subject of climate change in USA schools (https://edition.cnn.com/2017/06/14/health/climate-change-schools-partner/index.html). Although 100% of climate scientists agree that global warming is linked to the burning of fossil fuels, many middle and high school teachers apparently are not aware of this consensus [1]. This disconnect of climate knowledge between scientists and educators was captured by the National Center for Science Education, a nonprofit that works to promote science over ideology. Their survey found that relatively few teachers had even a college course that devoted as much as a single class to climate change and that many teachers present misinformation about climate change or avoid teaching it entirely. Also, Gerald Lieberman notes that it is really inconsistent in terms of how climate change is being taught. He is director of the California State Education and Environment Roundtable, which works closely with the California Department of Education on instructional strategies related to the environment.

CNN reports on a positive note, 18 states and the District of Columbia have approved the Next Generation Science Standards, which were developed with the help of several national science organizations and unveiled in 2013 (https://edition.cnn.com/2017/06/14/health/climate-change-schools-partner/index.html). A critique of these standards is that climate change instruction is largely relegated to earth and general life science classes, not so much to standard biology, chemistry and physics, the science subjects that most students focus on in high school and middle school. However, some Advanced Placement environmental science and environmental chemistry classes are presently being taught climate change, such as at Woodside High School in California. This school also is working on weaving climate change into the biology, chemistry and physics curricula, taking it further than the Next Generation Science guidelines suggest.

California recently just adopted the Next Generation Science Standards, but some educators say it may take seven to 10 years for California to fully execute the standards and a decade or two for the rest of the nation to follow suit because these are gigantic educational systems. Fortunately, some school districts, like Woodside High School, may adopt or innovate their own standards before their own state legislatures do. For example, in Wisconsin, the vast majority of school districts seem to have adopted the Next Generation Science Standards, even though the state hasn't done so. In the end, educators on the front line note that state science standards do make a difference, because they influence what's in textbooks, local curricula, and statewide tests.

As CNN notes, at the end of the day, when the classroom door closes, it's really going to be the individual teacher who determines whether or not global warming and human-caused global change is going to be properly presented (https://edition.cnn.com/2017/06/14/health/climate-change-schools-partner/index.html). A teacher in the rural Texas Hill Country provides an excellent example of how to present climate change. As one might expect in this conservative bastion of the nation, many of the students at the beginning of the semester say climate change is all lies or fake news. To counter this, the teacher adopts a "show, don't tell" approach when teaching his students about climate change. He has the students track data from USA sources like NOAA, USGS and NASA. Rather than telling the students, they find out themselves that any reliable data source they choose will show that global warming is real and they change from climate change denial to climate change believers by the end of the class term.

I have just given an excellent example of how a teacher in Texas approached the teaching of climate change in an ideal way. Unfortunately, this is just the tip of the iceberg because as I have noted climate change is only one of the many human-caused global changes that are taking place on our planet. Not only is there a present lack of teaching about climate change, but I also suspect that few USA grade or high schools address the totality of human-caused global changes and the imminent danger of the combination of these changes occurring at the same time. Even the US interagency program, National Global Change Research Plan 2012–2021, only focuses on climate change and not the other global changes in air, water and soil quality (https://www.globalchange.gov). Global change also includes depletion of water, soil, fisheries, coral reefs and coastal habitats. As a result of these realizations, I think there is now an important need for education about all types of global change and for a major revision in our science education for the United States public schools and other global school systems as well.

From my experience, we have had a lack of teaching about natural science in the United States as opposed to hard science. Part of this problem is because science education has mainly been focused on the hard sciences of physics, math and chemistry, such as the new implementation of the STEM (Science, Technology, Engineering, and Mathematics) program in USA (https://www.ed.gov/stem). There also is a focus on science training in biology, but often the class material concentrates on cellular biology, genetics, invertebrate and vertebrate zoology at the expense of ecology and environmental biology. I doubt that many students had the unusual experience that I had in the 1950s, where our biology classes collected field data each year on the pheasant populations near Mankato, Minnesota. From these data we realized that humans were destroying the habitat and the pheasant population was continuously declining and how now pheasants are nearly absent in this part of southern Minnesota.

We can see the results of this educational failure when looking at some of the dismal statistics of the United States natural science education. For example, Pew research in 2009 showed that only 32% of the USA population believes in natural evolution and 38% think that creative design should be taught as part of the science education, including former President George Bush Jr. It was thought provoking while living in Texas, when a good friend from Columbia one day said, I come from a developing world, Catholic country and we believe in evolution, but I do not understand why most people in Texas do not believe in evolution.

In my experience with my daughter's education in Spain and mine in the United States, I can see part of the reason for the failure of natural science education in the United States. Previously in USA high schools, it often has been normal to only have about one or two science classes per year, and mathematic's classes. As a result, the majority of USA high school students may end up taking as few as three or four science classes by the time they graduate, even when they have an interest in science.

In contrast in Spain, my daughter taking the high school science lane from 2003 to 2007, had four years of every science including mathematics, biology, chemistry, physics, and earth science. My daughter's education in Spain included discussions of climate change and other global change in an extra course that she took on environmental science. In this course they discussed aspects of air, water and soil pollution. My other daughter a year older was in a social science lane in high school and took fewer courses in science mathematics, biology and did not have any teaching of global warming or other change. The reason, that my daughters in Spain had so much science education, is the Spanish high school students typically take 12 or 13 courses per term, versus the typical five or six courses per term that United States high school students take. This comparison comes with the caveate, that before high school, USA students outperform Spanish students in general science (https://www.oecd.org/pisa/PISA-results_ENGLISH.png). However in high school, Spanish students that finish a science lane appear to have a broader education in natural science than USA students.

On the brighter side, there now appears to be a rapid improvement in more innovative schools in the United States, so that global warming and even some global changes are taught in grade and high schools. The local high school, where I previously lived in California, now has an advanced placement class in environmental science where global warming is taught. Also during the past few years, my five granddaughters and grandsons have had environmental science classes covering global change in their Oregon, Virginia and even Uganda schools. For example, my 16 year-old grandaughter at an international school in Uganda had a junior year unit on global and climate change that includes research disproving denial data. Although global change is still not well covered in many USA high schools, the university students majoring in earth, biological and especially environmental science do get information about global changes related to their science specialty. This is important because many of them will become teachers or professors and they will now be trained to teach all aspects of global change.

The most encouraging thing of all in the USA is that global change topics are even reaching down into some grade and junior high schools. A friend in California is now teaching effects of human impacts, and global change in her 6th grade earth science class. My granddaughter in Virginia had some teaching about climate change in her junior high school and a grandson in Oregon took a course in oceanography in junior high school. I recently have talked to some other grade and high school children from other parts of the USA and they also have been taught aspects of environmental science and global change.

The education about global warming also has improved in Europe. For example, my wife and her doctoral students teach special earth sciences courses to the grade school children in Granada, Her Greek student Dimitris, who helps teach, reports that the 10 year-old Granada students are knowledgeable about global warming. He also notes, that similar to my daughter in Spain, the science students of his generation were educated about global warming, but that social science students did not learn about global warming. Fortunately, now Dimitris says this has changed in Greece and all children get an education about global warming and are knowledgeable because of all the media coverage.

My anecdotal experience suggests that there has been a general lack of teaching earth science in the United States high schools, where it would be most logical to teach climate change and the impact of other natural and human-caused global changes. A good earth science course should cover atmospheric, ocean, earth and environmental science. With this combination of science backgrounds, high school students could understand the problems of climate and global change. My older daughters in the 1980s took earth science classes in the Palo Alto high school that they attended. Several times I gave lectures to the students in my daughter's class. However, since that time, the school has dropped the earth science courses. As a result, the many USGS earth scientists in the area rarely provide special lectures and background for these students. This Palo Alto high school system, usually rated as one of the top ten in the USA, apparently has eliminated the earth science courses in favor of more advanced STEM courses in the hard sciences of physics, chemistry and mathematics, although some environmental science courses have been substituted.

Even where earth science and geoscience courses (geology, earth, ocean, atmospheric sciences) continue in school systems, a new summary of the present state of geoscience education in the United States shows that human-caused global change topics generally are not taught [2]. The new study confirms my concerns about educating grade school, high school and introductory university students about global change. Education at these levels is the only school exposure to global change problems that most people will get. Even more critical is that for future grade and high school teachers, an introductory university earth science course is the only place where they may encounter formal training about global change problems. Teaching about global change in geoscience courses is important for the millions of students taking these courses because this training will be critical to make personal, professional and societal decisions for addressing global change problems.

The analysis of geoscience courses in USA provides both good news and bad news about education for global change. The good news is that a new set of science standards has been developed by the National Research Council (NRC) for grade and high school science education and these standards emphasize earth and human activity [2, 3]. These new NRC standards have provided guidance for the development of the Next Generation Science Standards (NGSS) (https://www.nextgenscience.org/). The NGSS standards are being implemented by multiple states, with each state developing its own science guidelines, but there is no overall federal mandated science program. The NGSS program takes a systems approach and emphasizes the dynamic global change interactions between the atmosphere ocean, land, and life [4] The other good news, as I have described previously, is that teachers and professors have access to many new research-based instructional resources (e.g. the Internet, media) and new curricular materials that emphasize systems thinking and connections between Earth and human activity).

The bad news is that these new set of NGSS educational standards emphasizing global change have been poorly implemented in geoscience courses, especially for university level geology courses. The most common topics listed in course descriptions suggest that an Earth Systems approach is not widespread (e.g. only 4% of the geology courses) and human interactions with Earth are not emphasized [2]. Almost none of the top fifteen topics in course descriptions for earth science, geology, or oceans highlight connections to humans. Only 6% of the geoscience course descriptions focused on Earth systems, but not ecosystems, and only 6% included a learning outcome that referred to human activity and/or society. Few courses engage students in working on local problems, environmental issues, or problems of global or national interest, but those who teach oceanography and Earth science are more likely to do so than those who teach geology. In summary, human interactions with Earth are not emphasized in most introductory geoscience courses.

There are a number of solutions to the poor present instruction of global change problems in geoscience courses. There should be an Earth-Systems teaching approach that includes the use of real-world global change problems that integrate multiple disciplines [2]. There also should be model-based reasoning to understand complex systems over time, such as climate change and human interactions with the global changes on the Earth. Also connections to individuals and societies should be emphasized in introductory geoscience courses. Investigating phenomena that connect to their lives and communities can show the relevance of a geoscience course, provide students with motivation for learning and positively impact recruitment and retention of students into the geosciences.

7.2   Revolution in Education

From the Sect. 7.1 evidence of minimal teaching of global change in many of the USA and worldwide educational systems, here is a plan that I suggest for a new system. This plan includes: (A) major changes to middle and high school science curriculums and science course requirements, (B) significant increased training in the natural sciences with a focus on all the human-caused global changes, (C) a new focus on educating political and government officials about the multiple problems of global change (e.g. Figs. 4.4, 6.1 and 6.2), and (D) education in political and business ethics so that there is a focus on preserving a sustainable Earth for the continued survival of the human race.

In our middle schools and high schools, we need to continue our good progress with the hard sciences in programs like STEM, but we also need to add and give equal weight to the natural sciences in a program that I call SEE (Sustainable Earth Education). A way to do both is provided by the example of high school students in Spain, where they make a commitment for the last two years to focus on a science curriculum or other non-science curriculums. Those in the science curriculum continue with further STEM hard science courses, but they also continue with a SEE natural science curriculum as well.

For high school students in any country that do not intend to have a career or increased background in the STEM sciences, they should be required to take at least two year-long SEE natural science courses. These natural science courses should consist of earth and environmental science. For SEE the first year, an earth science course should be taken that includes science of the atmosphere, ocean, soil as well as onland surface and groundwater.

With the background in the earth sciences and the general environments of the earth, students in SEE the second year should take a course on environmental science. The second-year environmental science course should include the fundamental knowledge of limiting factors, carrying capacity and the sustainability of a living population. Students should learn that a negative change of carrying capacity and only one limiting factor could reduce or terminate a living biological population. The second year students can then combine this knowledge of limiting factors with knowledge of human caused global changes in the Earth's different environments. With this combined knowledge, these students will be aware of the severe threat for a sustainable human population because of the multiple global changes that are destroying our air, water and soil quality, which are necessary to sustain human life (e.g. Figs. 4.4, 6.1 and 6.2). The discussions of the types of global change, that are described in Chaps. 4 and 5, also need to focus on the fact that there are multiple global changes taking place at the same time and becoming limiting factors to human life, perhaps even in less than a century.

In addition to human-caused climate and other global changes, the second year course also should include natural global changes related to earthquakes, volcanoes, and weather disasters. With the first-year course in earth science and the second year course in environmental science, students would understand the need to address the problems of both natural and human-caused global change, which result in a huge cost in lives and economic losses every year.

With knowledge of these global change losses, in the future as citizens, these students with a general background in earth and environmental science will be aware of the decisions and needs for mitigating these natural hazards and human-caused global change. For example, the USA Pacific Northwest is faced with eventual catastrophic losses similar to those from the 2004 Sumatra and the 2011 Japanese earthquakes. With a good understanding of earth science, the general population will realize the need for spending on infrastructure to prevent as much loss of life and property as possible from natural hazards. Then maybe there will be progress to retrofit buildings and the 1700 bridges for earthquake safety in the Pacific Northwest.

With regard to teacher training, all grade school, middle school and high school science teachers should take SEE courses that include earth and environmental science. Science teachers with this background can then cover the topics of all human caused global changes as well as global warming on which there has been the main focus up to now (e.g. Figs. 4.4, 6.1 and 6.2). As described in my previous example of an excellent teaching style, students should investigate reliable scientific resources of data themselves to become convinced of the aspects of global change. The students nowadays, and even children a few years old, learn how to look up information on the Internet, so they are used to finding information in this manner. Teachers can guide the students to the different scientific data sources that they can use for their own investigations. One further idea is that different students or groups of students could focus on a specific type of global change and then this student or group of students could report on their findings to the class. With this method, student-peers would have the most effective way to convince other students of their findings about global change.

Equally important for the general population, knowledge transfer about human caused global changes should take place with politicians and government officials. At present, it appears that local politicians and government officials are becoming more aware of global warming and its consequences. It is difficult to ignore the catastrophic Atlantic hurricanes, flooding and wildfires from 2017 to 2020 in North America and Austrialia. In certain areas, the local politicians and officials are aware of other global changes, such as when there has been severe pollution of local land and water from chemical manufacturing. However even with the local politicians and officials, there probably is not an awareness of the broad spectrum of global changes throughout the Earth and the potential threat to the sustainability of the human population (e.g. Figs. 4.4, 6.1 and 6.2).

The change in SEE school curriculums and increased teaching of natural science will take a number of years to educate voters to influence the national level of politicians and government officials. Consequently, there is an urgent need for scientists to brief the Congress, executive branch and staffers about global changes and the immediate need to execute government regulations and infrastructure to sustain the Earth's human population.

Most important of all and certainly most difficult of all, is the need to change consumer habits as well as the political and business ethics. There must be an understanding that there cannot be business as usual with the unrestrained use and continued pollution of the Earth's air water and land resources (e.g. Figs. 4.4, 6.1 and 6.2). We must develop a political and business will to take action to provide for a sustainable human population. Taking such action actually will result in a win–win situation because developing the regulations and infrastructure will create many jobs. This is already happening in China, whereas the recent avoiding of action from 2017 to 2020, such as leaving the Paris Agreement, is having the opposite effect by moving these jobs to China rather than keeping them in the United States.

The way forward to change the ethics of government and business leaders, is to become an educated public about global change and then applying pressure to one's political representatives and having stockholders apply pressure to business leaders. The latter has recently started to become effective in forcing companies to take steps to alleviate problems of global change, such as acknowledging global warming and developing measures for conservation of energy. Again, something like energy conservation is a win–win situation where the Earth wins and the company saves money.

7.3   Saving Our Planet

Everyone needs to be stewards of our global environment and work for future sustainability of human life on earth. This requires factual scientific education for a sustainable Earth from an early age on. We need education to prevent irresponsible denial and skepticism of proven scientific evidence, such as denial of global warming and recently of Covid-19 by populist leaders. For example, in contrast to scientific evidence, global warming has become a political issue in USA, because it benefits conservative politicians and companies for economic gain. As a result of propaganda by fossil fuel companies, until recently the belief in global warming decreased 10 to 20% and in some polls only half of the United States population believed in global warming (https://thebridge.agu.org/2013/06/05/global-warming-public-opinion-and-policy/) [20].

Even worse for me as a scientist, some polls show the effectiveness of climate change denial funding in that barely more than a third of the USA population in 2019 has confidence in climate change scientists. As a result, between 2017 and 2020 many laws and regulations have been changed in USA. Consequently, CO_2 and CH_4 emissions, global warming and environmental contamination are continuing, which has potential catastrophic results for a sustainable earth and its human population. The Earth will survive because it has been here for billions of years throughout warmer and colder climates (e.g. Fig. 4.5). However, we humans need a sustainable planet on which to survive. Whether the earth will remain sustainable for humans is unknown, but at a minimum with present rates of environmental degradation, the quality of life will be diminished for the humans that survive (Figs. 4.4 and 6.2).

Even if you are skeptical that there is human caused global warming, no one can deny that other human-caused global changes are taking place (e.g. Figs. 4.4, 6.1 and 6.2). Everyone can see the pollution of air, water and soil in India or China where the populations are largest. Already because of global warming, loss of habitat and pollution, the present rate of animal and plant species extinction is approaching the rate caused by a meteorite hitting the earth about 66 million years ago (e.g. extinction of the dinosaurs) (Fig. 4.4f). We humans can observe alreaady that there is a significant loss of ground water (e.g. India), ocean fisheries, coral reefs, mangrove forests and farmland because of pollution (e.g. China) desertification (e.g. raining mud in Spain) and erosion (Figs. 4.4c and 6.2 Marine fish capture, Terrestrial biosphere degradation, 6.5) [5].

A reasonable course for the long term is to put our efforts into A) regulations reducing the input of CO_2 and CH_4 plus other pollutants into the air, water and soil, B) conservation of water and soil, C) infrastructure that provides renewable energy, D) avoiding unnecessary consumption and food waste, and E) education of students, general population, and political leaders on how to maintain a sustainable Earth. As a scientist, it is always astounded me why there are not positive increases in government budgets to have a sustainable planet and human population and only budget increases for weapons to eliminate the human population. A possible way to unify the USA is to develop a new type of Marshall Plan to save the earth by exporting knowledge and technology instead of exporting weapons. We could create an Earth Corps like the Peace Corps to help countries to develop sustainable economies and environments. Creating a sustainable Earth will provide a much more secure planet than developing more nuclear weapons. These changes in political will must begin now, because we cannot wait for an environmental crisis to occur when the Earth has passed an environmental threshold and it is impossible to make the changes for a sustainable human population on our planet.

7.4   Sustainable Earth Take Home Points

(1) Human-caused global change includes climate warming, and all other human-caused effects on the Earth's air, water and soil that sustain human life (e.g. Figs. 4.4, 6.1 and 6.2)

(2) Global warming is already taking place as shown by the patterns of increasing CO_2 and CH_4 in the air, accelerating warming of the atmosphere, a consistently warming ocean and rising sea level, melting glaciers, and worse extreme weather events (e.g. hurricanes, floods, droughts, fires, heat waves, blizzards) (Figs. 4.1, 4.2, 4.3, 4.4g, h, 4.6, 4.7, 4.8, 4.9, 4.10, 4.11, 4.12, 4.13 and 4.14).

(3) Global warming effects in 2017 to 2020 alone have cost thousands of lives and billions of dollars in damage, which will continue into the future (Figs. 4.12 and 4.13).

(4) Eventually eliminating burning fossil fuels not only reduces global warming, but also atmospheric pollutants that cause about 25% of global deaths.

(5) Developing a renewable energy supply and reducing CO_2 and CH_4 in the atmosphere creates millions of jobs as shown by China which

already has 3.5 million new jobs that are expected to increase to over 10 million jobs.

(6) In contrast, the climate change denial culture of the USA has resulted in only a few hundred thousand jobs and a lost opportunity for millions of new jobs in a major new industry.

(7) Increased conservation of energy, use of renewable energy supply and more energy efficient transportation such as trains and hybrid/electrical vehicles will help elevate the global population above the poverty line.

(8) Air pollution kills an estimated 9 million humans a year, as many as 1 in 4 in low and middle income countries and is responsible for about a third of lung cancer, pulmonary and heart disease in the world (e.g. Figs. 4.13 and 6.2 Nitrous oxide, Methane)

(9) Clean water is fundamental to support life and the supply needs to be increased through conservation and new technologies for the growing population, and to prevent the millions of lives, particularly of children, that are dying presently from polluted water (Figs. 4.4b and 6.1 Population, Water use).

(10) To preserve farmland, conservation measures are required. Pollution from industrial wastes needs to be prevented and global warming has to be slowed to reduce droughts, wildfires and desertification (Figs. 4.1, 4.8, 4.13 and 6.2 Terrestrial biosphere degradation).

(11) Optimizing fertilizer use on farms by applying nutrients at the appropriate rate, in the right place, at the right time, can reduce overall fertilizer application emissions while also enhancing water quality and decreasing dead zones in the ocean (Figs. 6.1 Fertilizer consumption, 6.2 Nitrogen to coastal zone) [6].

(12) Reducing meat consumption, and eliminating use of antibiotics in animals, benefits humans and the environment (Fig. 4.4i).

(13) Former and current communist countries have been an example where the lack of regulation and combination of air, water and soil pollution has resulted in higher mortality rates (e.g. https://www.sciencedaily.com/releases/2017/11/171106100658.htm)

(14) Radioactive contamination affects air, water and soil, and other than nuclear war, the severest problem is finding permanent storage of high level wastes, which 1 in 3 USA Americans live near.

(15) Worldwide sustainable fisheries and marine sanctuaries need to be established to provide continuing food supplies and preserve marine habitats (Fig. 4.4c and 6.2 Marine fish capture, Shrimp aquaculture).

(16) The ocean pollution of plastics, oil tanker cleaning wastes, other trash and fertilizer runoff into the oceans also must cease to preserve sustainable fisheries and the ocean habitat. The use of onetime only plastic containers needs to be reduced and eventually eliminated by countries and companies (Figs. 4.4c and 6.1 Fertilizer consumption, 6.2 Nitrogen to coastal zone).

(17) Population growth plus consumer consumption driven by economic goals is the fundamental cause of all these human-caused global change problems of burning fossil fuels, global warming, pollution of air, water and soil, over fishing and destruction of major habitats like coastal wetlands, mangrove forests and coral reefs (e.g. Figs. 4.4, 6.1 and 6.2).

(18) The best solutions for human-caused global change problems are (A) conservation of air, water and soil resources, (B) prevention of pollution, (C) renewable energy supplies, (D) raising humans out of poverty, (E) political will from widespread awareness and education (F) limiting consumerism to sustainable levels and (G) most of all evolution to a sustainable population for the earth (Fig. 6.6).

(19) The general public and politicians need to have education about all types of global change. The public and political leaders then should institute regulations for pollution, land use, and renewable energy. Corporations and political leaders should make the Earth's and human health the top priority and utilize scientific facts rather than fake news for economic gain for only a few.

(20) The world's education systems need to be revolutionized so that environmental sciences are given equal standing to the hard sciences of mathematics, chemistry and physics. In this way, the general population will have the necessary knowledge about the combined effects of human-caused global changes and will be able to make wise political choices (e.g. Figs. 4.4, 6.1 and 6.2).

(21) The education of political leaders about human-caused and natural global changes is critical to avoid crossing environmental thresholds, which makes it impossible to make the changes for a sustainable human population on our planet. The knowledge about natural hazards needs to be utilized to warn humans about dangers and to implement necessary infrastructure to reduce human fatalities and economic losses.

(22) In summary, humans have no other planet to live on or move to. We now sacrifice the long-term sustainability of our Earth for human habitation in exchange for short-term economic gain.

7.5   Solutions for Global Change

My family and friends sometimes think of me as Dr. Gloom and Doom, because of my constant reference to the problems of global change. As a reader this may be the impression left by my book, because of the long list of global change problems that I have described. I have summarized the global change problems to make the public aware of what we are doing to the health of a sustainable Earth and human population. However, I agree with journalists and psychologists, that constantly framing global change problems in terms of immediate disaster [6], can lead to people feeling hopeless and that there is nothing that can be done to prevent global change problems. Consequently, I would like to leave the reader with the long list of the positive changes that the public and global governments have accomplished to combat global change problems and how this provides hope for the future of a sustainable Earth.

These are some of the positive global change solutions that the governments, public, private organizations and industry have made during my lifetime:

7.5.1   International Agreements

- The Antarctic Treaty System started in 1959 and eventually with 55 countries, was one of the first international agreements to protect a continental and oceanic environment.
- Since the Treaty, Antarctic scientists from countries with antagonistic politics such as the cold war have worked together harmoniously to set a good example for other people globally.
- The Partial Nuclear Test Ban Treaty in 1963 banned atmospheric testing to help prevent global radioactive contamination.
- Since 1968, a ship for academic deep-ocean drilling has been supported internationally by 22 countries as the Deep Sea Drilling Program, Ocean Drilling Program and Integrated Ocean Drilling Program. This program has provided deep ocean sediment records of global change and particularly climate history for over 100 million of years.
- These ocean drilling studies have given us perspective on past climates, which we can compare for present and future climate change.
- The Law of the Sea signed by 157 countries in 1982 defined territorial and international waters, and helped to determine the areas of responsibility for global change problems in the ocean.

- Green political parties, which are concerned with global change problems, have formed during the past few decades in some countries of the world such as in Europe and New Zealand.
- In 1987, 197 countries signed the Montreal Protocol on substances that deplete the ozone layer. It already has reduced the size of the ozone hole and may restore the ozone layer by the end of the twenty-first century (Fig. 4.4a) (https://weather.com/science/environment/news/ozone-hole-closing-nasa).
- The Paris Agreement of 2015, to reduce CO_2 emissions, has been signed by almost all of the world's governments, and other countries are taking up the leadership abandoned by the USA since its withdrawal (Figs. 4.2, 4.4g and 4.15).
- The United Nations Sustainable Development Goals (SDGs), also known as the Global Goals, were adopted by all United Nations Member States in 2015 as a universal call to action to end poverty, protect the Earth and ensure that all people enjoy peace and prosperity by 2030.
- In 2016 the Port State Measures Agreement (PSMA) was put in force by 55 countries to combat illegal, unregulated, and unreported (IUU) fishing.
- In 2018, 18 nations at the second Arctic Science Ministerial Meeting and representatives from Arctic indigenous groups called for increased efforts for international scientific collaboration because of the warming trend in the Arctic (Figs. 3.9, 3.10, 4.8 and 4.9) [7].
- The International Monetary Fund reports that the world's countries spend 5.3 trillion dollars a year in fuel subsidies, which could be used to fund renewable energy.
- After the near extinction of many bird species at the top of the food chain, DDT use has been nearly eliminated by most countries. Many of the nearly extinct species have returned to near normal populations, for example eagles, perigrine falcons, and brown pelicans.
- Lead has been mainly eliminated from gasoline and paints in the developed world.
- Mercury use has been reduced globally in many products such as paints where it was added to prevent mildew.
- Sulfide emissions of power plants and acidification of lakes has been greatly reduced.
- Solutions for a sustainable potable water supply are being implemented globally through better water management, conservation, new technology like drip irrigation, recycling of waste water, desalinization, and better selection of crops for arid regions (Fig. 4.4b).

- The populations of most developed countries are decreasing or have become stabilized so that these countries that use the most resources per capita are contributing less to global change problems (Fig. 6.6).
- The International Whaling Commission has mainly eliminated whale hunting except for Japan.
- Because of regulation, the Pacific gray whale population in the Pacific Ocean has recovered from a few hundred at the beginning of the 20th century to a population of approximately 22,000 by the beginning of the twenty-first century [8].
- Although more than half of the world's coral have been lost in the past three decades, there are successful efforts to replant corals in the Caribbean by growing 50,000 corals in nurseries and replanting more than 10,000 corals [9].
- There is some evidence of the poleward range expansion of tropical coral reef associated organisms since the 1930s because of warming oceans, and the speed of these expansions has reached up to 14 km/year in the Japan Sea (Yamano, 2011).
- Some boreal peatlands, which make up less than 5% of the total surface area of the planet but contain roughly 30% of all global soil carbon, may prove more resilient to climate change than previously thought and could thrive as major carbon sinks even as northern regions become drier and less hospitable [10].
- Global energy demand is expected to plateau after 2035, despite expansion of population and economic growth, and renewable energy is expected to make up 50% of electrical power generation by then [11].
- Within 5 years, there is projected to be increased use of renewable energy as the cost decreases and is competitive with the fuel cost of existing conventional power plants [11].
- As the cost of batteries continues to decline within the next 5–10 years, many countries will reach the point at which electric vehicles are more economic than internal combustion engines for passenger cars and most trucks [11].
- Despite continued economic and population growth, because of a drop in global coal demand and flattening oil demand, worldwide CO_2 emissions are expected to start to decline by the mid-2020s [11].
- CO_2 emissions per person have already decreased nearly 5% in USA Europe and Australia since 2005 (The Economist October 19, 2019).
- Although raising livestock provides about one third of the unnatural sources of CH_4, there is a worldwide movement to reduce eating meat.

- At the 2015 Paris Agreement meetings a number of countries began a plan to reform agribusiness to be more sustainable [12].

7.5.2   National and Local Agreements

- To clean up the deadly London smog, the United Kingdom passed the Clean Air Acts in 1956 and 1968.
- The USA formed the Environmental Protection Agency (EPA) in 1970, passed the Clean Water Act in 1972 and passed the Clean Air Amendments in 1990, which reduced the ozone layer, acid rain and toxic pollutants, as well as improving air quality and visibility (Fig. 4.4a).
- In 1990 The USA congress made an amendment to the EPA Clean Air Act to develop a cap and trade system to reduce sulphide SO_2 emissions and limit acid rain in lakes. Between 2005 and 2015, the SO_2 emissions in the eastern United States were cut by 80%. (https://earthobservatory.nasa.gov/images/87182/sulfur-dioxide-down-over-the-united-states)
- In 1993 the California Regional Clean Air Act established a cap and trade system to reduce CO_2 emissions (Fig. 4.4g).
- British Columbia developed a carbon tax to cut fuel consumption and CO_2 emissions without harming economic growth (Fig. 4.4g).
- Canada implemented regulations reducing CH_4 emissions in 2020, which prior to that caused 26% of Canada's total greenhouse gas emissions.   (https://www.canada.ca/en/environment-climate-change/services/canadian-environmental-protection-act-registry/proposed-methane-regulations-additional-information.html).
- Many USA federal government agencies are cooperating and coordinating research in the US interagency program, National Global Change Research Plan 2012–2021. However, the programs main focus on climate change needs to be broadened to all aspects of global change in air, water and soil quality (e.g. Figs. 4.4b, d, e, 6.1 and 6.2).
- In October 2015 some former defense secretaries, secretaries of state, and 40 senators (Republican and Democrat), military commanders and national security experts, published a full-page letter in the Wall Street Journal saying that global warming is causing a world that is more unstable, resource-constrained, violent, and disaster-prone (e.g. Figs. 4.1–4.14) [13].
- A new USA federal government Interagency Collaborative for Environmental Modeling and Monitoring began in 2018.
- To reduce the extreme forest fires, in 2018 the USA congress passed the Healthy Forest Restoration Act, which also helps to develop sustainable forests that reduce global warming by absorbing CO_2 (Fig. 4.14).

- The U.S. Department of Defense is preparing for global warming building infrastructure for higher sea level and preparing security measures for the altering geopolitical relations that have helped to cause armed conflicts such as Rwanda and Syria [13].
- Many USA states and cities continued to abide by the Paris Agreement during 2017 to 2020 and reduced CO_2 emissions, and the USA rejoined the Paris Agreement as well as WHO in 2021.
- After severe air pollution problems, India and China with the world's largest populations are taking steps to reduce coal use and CO_2 emissions . They have the opportunity to leap frog over fossil fuel use to renewable energy from solar and wind power.
- Both India and China are already planning to have only electric cars within a few decades.
- Some European countries already have 100% renewable power and others have 50% or more.
- The USA has established sustainable ocean fisheries and some countries are following this lead (Fig. 6.2 Marine fish capture).
- Several countries have established marine sanctuaries to protect habitats and marine life as well as sustain fisheries.
- Loss of natural fisheries has been augmented by fish farming, which has become more efficient and producing healthier farm fish (Figs. 4.4c, 2.3 Shrimp aquaculture).
- A number of countries joined a United Nations Blue Carbon Initiative to protect coastal environments such as mangrove forests, marshlands and estuaries (https://www.thebluecarboninitiative.org/about-the-blue-carbon-initiative)
- The USA and other countries began marshland restoration from areas that previously had been utilized for urban development and commercial purposes such as salt ponds in San Francisco Bay.
- The Mediterranean countries cooperated to reduce pollution in the Mediterranean Sea (europa.eu/world/agreements/prepareCreateTreaties Workspace/treatiesGeneralData.do?step = 0&redirect = true&treatyId = 487)
- Governments and private NGOs began a campaign to reduce the use of plastics and in 2018 private groups started to clean trash accumulations from the ocean.
- Great Britain and the European Union have passed a directive banning single use plastic containers and some USA grocery companies are also taking steps to eliminate such containers (https://ec.europa.eu/commis sion/presscorner/detail/en/IP_19_2631).

- Industry and governments have been cooperating to remove micro- and nano-plastics from products.
- Private NGOs and government assistance programs are helping developing countries to conserve water and develop clean water sources.
- The California Sustainable Groundwater Act has now organized stakeholders to plan for sustainable groundwater use.
- The U.S. Army Corps of Engineers and other worldwide civil engineering groups have initiated Nature Based long-term engineering to restore coastal areas from urban development, hurricane storm damage and rising sea level.
- In Canada's Western Arctic, native communities are adapting to higher temperatures and earlier snowmelt by modifying their hunting strategies (e.g. Fig. 4.8). Because melting permafrost destroyed access routes to hunting grounds, community members have created new trails, identified new hunting grounds and targeted different species.
- The Green New Deal proposed by U.S. Representative Alexandria Ocasio-Cortez addresses two urgent problems: global warming and poverty. Well-paying green jobs (e.g. creating solar panels, wind turbines, smart batteries) would be good for the Earth and help lift people out of poverty.
- The Union of Concerned Scientists estimates that a standard requiring that 25 percent of all electricity come from renewable energy sources by 2025 would create nearly 300,000 new U.S. jobs, far more new jobs than would result if we relied on coal and natural gas.
- New York Times notes that USA CO_2 emissions could be reduced 50% by 2050, through applications of presently available solutions such as California and New York states regulations requiring future zero carbon source power generation, Norway's electrical vehicle requirement, British Columbia's present carbon tax, Canada's plan to reduce methane from oil and gas operations, and European Community legislation to eliminate ozone depleting chemicals.
- The USA Senate Committee on Commerce, Science, and Transportation on November 13, 2019 approved the bipartisan Save Our Seas 2.0 legislation to deal with ocean plastics pollution and builds on a law about ocean debris passed in November, 2018.
- In 2019 Senator Merkeley of Oregon introduced bill SB 1691 Wildfire-Resilient Communities Act to fund forest fuel reduction and reduce the chance of catastrophic wildfires in western USA.
- The Nature Conservancy already is undertaking wildfire suppression work in Oregon to create climate-resilient forests that store more carbon.

- The Nature Conservancy also is restoring coastal wetlands, planting trees and preventing grassland tilling in Oregon to keep carbon locked in the ground.
- Although growing population has put ever more people at risk, Bangladesh has adapted to global warming with a cyclone early-warning system, concrete shelters, and sea walls so that the 2020 Cyclone Amphan caused only about 100 deaths whereas similar storms early this century killed thousands to hundreds of thousands.
- An estimated 15,000 died in France in 2003 as a result of the August heat-wave, whereas in 2019 a similar event is estimated to have killed only 1,500 because of adaption measures to global warming.

7.5.3 Technology and Education

- About two-thirds of the world's population lives in countries where renewable energy represents the cheapest source of new power generation (BloombergNEF https://about.bnef.com/new-energy-outlook/).
- The world's best solar power schemes now offer the cheapest electricity in history, with the technology cheaper than coal and gas in most major countries.(https://webstore.iea.org/worldenergy-outlook-2020).
- Wind and solar renewable energy are predicted to make up almost 50% of world electrical power generation in 2050 and help put the power sector on track for only 2 degrees warming until at least 2030 (BloombergNEF https://about.bnef.com/new-energy-outlook/).
- Renewable energy will meet 68% of Spain's electricity demand by 2030 and almost 90% in 2050, up from 40% today (https://www.acciona.com/pressroom/news/2019/december/renewables-generate-almost-70-spain-s-electricity-2030-says-joint-acciona-bloomberg-nef-report/).
- In carbon storage experiments tied to geothermal power plants in Iceland, 90% of injected carbon dioxide (CO_2) transformed into minerals in just 2 years. Standard carbon storage methods can take thousands of years to do the same [14].
- The green revolution of the 1970s greatly increased food production so that the present world population can be sustained (https://en.wikipedia.org/wiki/Green_Revolution).
- Korea has made it illegal to send food waste to landfills, which reduces methane emissions, creates fertilizer and animal feed, plus saves billions of dollars (New Yorker, March 2020). This method could be applied world-wide and in the USA alone would reduce greenhouse gas emissions by the equivalent of thirty-seven million cars on the road each year.

- Water resources are being improved by new techniques of desalinization, drip irrigation, and wastewater reclamation.
- New mathematical models (or fingerprinting) can determine if worse extreme weather most likely is the result of global warming. This method helps to differentiate between natural climate change and global warming affects [15].
- For example, the extreme warm temperatures of Siberia in 2020 are 600 times more likely to be related to global warming rather than to natural temperatures (CNN online news, July 16, 2020).
- Fingerprinting also indicates that the rise in global temperatures follows a pattern expected from greenhouse gases and not increased energy from the suns output (Newsweek, December 10, 2010).
- New Stormsensor technology can monitor stormwater runoff and help reduce the 32 trillion gallons of runoff in USA, which is the number one cause of urban pollution from chemicals in sewage (Nature Conservancy, summer 2019).
- New Watchtower Robotics technology can be used to find and locate water pipe leaks to help reduce the worldwide 20% water loss from pipe damage, which can provide more water for hundreds of millions of people (Nature Conservancy, summer 2019).
- New technology from This Fish enables seafool processing plants to digitize paper-based management systems, which will make the illegal and unsustainable fishing harder to hide (Nature Conservancy, summer 2019).
- Although 50% of high mountain glaciers will be melted by 2060, the Swiss are already developing new underground hydro electric systems to generate power from the lakes resulting from glacier melting.
- Earthquake engineering building codes for buildings and infrastructure have become well established for Japan, New Zealand and California, and are beginning to be implemented in other earthquake areas.
- Tsunami warning systems have been developed for the Pacific Ocean and are being developed for the Caribbean and Indian Oceans.
- Tsunami deposits are being mapped in many countries to assess hazards and plan for safe infrastructure (e.g. Goldfinger et al. [16]).
- Earthquake alert systems are operational in Mexico and Japan where high-speed trains are stopped within seconds of an earthquake.
- USA in 2018 has begun a new Shake Alert system to warn when an earthquake has occurred.
- The November, 2019 USA Interior Appropriations bill for programs includes $170.8 million for the U.S. Geological Survey's Natural Hazards

program for earthquakes, and includes $2 million for the ShakeAlert warning system.

- Paleoseismic history studies have been implemented in many countries to assess the frequency and strength of earthquakes and develop seismic risk maps and building codes for earthquake planning and infrastructure development (e.g. Fig. 3.21) (e.g. Goldfinger et al. [17]; Nelson et al. [18]; Polonia et al. [19]).
- Active faults that produce earthquakes are continually being mapped in many countries to assess hazards and plan sites for safe infrastructure such as nuclear power plants.
- Techniques have been developed to predict volcanic eruptions so that populations can be evacuated to safety and air traffic can be diverted from volcanic ash clouds.
- The U.S. Geological Survey and other international government agencies provide geologic hazards planning maps so that safe building and infrastructure sites can be developed in areas not subject to volcanic eruptions, landslides, unstable soils, floods and tsunamis.
- Natural hazard education has become widespread in the USA to protect the public from natural hazards of volcanic eruptions, earthquakes, tsunamis and extreme weather events.
- Prediction of extreme weather events such as tornadoes, hurricanes, floods, heat waves, super El Niños and unusual low temperatures has been greatly improved to warn populations and prepare emergency responders (Figs. 4.1, 4.12, 4.14).
- Grade school and high school students worldwide are being taught about global warming and other global changes (e.g. Figs. 4.4, 6.1 and 6.2). This will create a new generation of environmentally aware public.
- Numerous university and graduate students are pursuing degrees in environmental science, law, economics and engineering so that local and national governments can implement planning and regulations for global change.
- An encouraging sign of education for the global future are the two new polls showing that 72% of U.S. adults now believe the world is getting warmer and regard climate change as personally important, which is up 16% since March 2015 and 9% since March 2018 [20].
- Even more important, 62% of USA adults in these new polls now believe the global warming is caused by humans, which is up three times since 2011 and mainly results from their lives being impacted by extreme weather events [20].

- An encouraging sign of political education is that Alaskan Republican Senator Lisa Miskowski was the keynote speaker for the American Geophysical Union international meeting in December 2018 and gave a talk stressing the need for addressing climate change that has been ravaging Arctic Alaska (e.g. Figures 3.9, 3.10, 4.8).

7.5.4 Media Coverage

- The Internet is one of the most important sources for the general public to learn scientific facts about global change and solutions for a sustainable earth, although it also has been used to promote populist nationalism, climate change denial and influence democratic elections.
- In general, the media are providing some of the best information about global change and environmental protection, although some of the media promote climate change denial and less environmental regulation.
- Many newspapers and magazines provide excellent articles on global change problems and solutions, and you will see some examples of these that I have used for sources of information in this book (e.g. the Scientific American, National Geographic, The Economist, New York Times, Guardian, USA Today).
- International and local online news media provide up to date summaries of new environmental studies (e.g. CNN, BBC).
- Excellent summaries of global change problems are provided by documentaries such as BBC and Netflix (e.g. https://www.bbcearth.com/blu eplanet2/#; https://www.netflix.com/es-en/title/80049832, 80,216,393, 81,321,999).
- Government websites such as USGS provide detailed information on environmental studies and natural hazard events (e.g. volcanic eruptions, tsunamis, earthquake locations and magnitudes).
- Numerous international governmental agencies provide online information on environmental data (e.g. air pollution, water resources, water and soil contamination) (e.g. Figs. 4.4, 6.1 and 6.2).
- Excellent long and short term information on worldwide weather and climate effects (e.g. on land and ocean temperatures, sea-ice cover, glaciers, deforestation, desertification) are provided by websites of the National Aeronautics and Space Agency (NASA) and National Oceanographic and Atmospheric Agency (NOAA) who summarize data from satellites, ocean buoys, weather balloons etc. (e.g. Figs. 4.1, 4.2, 4.6, 4.7, 4.8 and 4.9).

- Private websites concerned with global change and environmental problems provide summaries of information (e.g. Union of Concerned Scientists, Environmental Defense Fund, World Wildlife Federation, Sierra Club, Greenpeace).

7.6   The Future for Humans

Predicting the future is an extremely dangerous exercise. However, I will give a best guess from my global experience and the present scientific data (e.g. Figs. 3.9, 3.10, 4.4, 6.1 and 6.2). The easy thing to predict is that humans will continue to face natural hazards such as earthquakes, volcanic eruptions, hurricanes, tornadoes, floods, forest fires, droughts etc. Natural extreme weather events will always occur. However, recent experience suggests that the climate related natural events are more extreme and cause more losses of life and infrastructure than they did previously (Figs. 4.12–4.14) (e.g. [21], Levi et al., 2019). Eventually in the future, the loss of life may be reduced, as humans become more aware of natural hazards and the extreme weather events while warnings and infrastructure continue to improve. Most likely the economic losses will continue to increase, because of the enormous infrastructure required to combat extreme weather events and populations moving into areas with extreme weather and natural hazards like earthquakes and tsunamis.

At present the outlook for reduction of CO_2 and CH_4 emissions, rising global temperature, and increased severity of extreme weather events is not promising (Figs. 4.1–4.15). The pledges and targets for the temperature rise, that were agreed to in the Paris Agreement, are actually nearly double those of the agreement goal and current policies are not even meeting these targets (Fig. 4.15). Even these pledges, that are being met, appear to be subject to creative accounting because CO_2 emissions for the transportation of imported consumer goods are not accounted for (The Economist, October 19, 2019). For example, air transport of flowers from the Netherlands to Great Britain results in 70 times more CO_2 emissions than ground transport and these emissions are not included in the domestic emissions budget of Great Britain. Even with these accounting problems, Great Britain has only half the CO_2 emissions per person compared to USA. A great reduction in CO_2 emissions would result if the USA, the world's second greatest producer of emissions, would reduce emissions per person to the level of Great Britain. Equally important is that China and India, the world's largest coal users, would reduce their use of it.

The effects of global warming are already present and will only get worse in the future when considering the above-mentioned lack of reduction in CO_2 and CH_4 emissions and global temperature rise (e.g. Figs. 4.1–14.14). The combination of population growth, limited resources and droughts has contributed to wars and genocide in the Middle East and Africa (e.g. Diamond [22]), and these conflicts most likely will increase. The Economist of March 2–8, 2019 reports that water conflicts have doubled between 2015 and 2019. Already these conflicts, extreme weather, and limited resources have resulted in climate refugees that have migrated by the millions to Europe and there will only be more in the future.

Even the USA has not been spared as the extreme hurricanes and wildfires have resulted in people moving out of the hurricane, flood and wildfire regions (Figs. 4.12 and 4.13). As a result of these migrations, populist, nationalistic and autocratic leaders and governments have emerged that refuse to acknowledge global change problems. For example in Brazil, deforestation of Amazon forests has doubled and these forests provide 20% of the atmospheres oxygen that humans need to breathe (CNN, July 2, 2019). The world's increase in populist nationalism is a concern, when what is needed to solve global change problems is internationalism and global cooperation, not everyone for themselves. The world will need bold new leaders with a global vision for the future.

Returning to my original lessons about environmental limiting factors and carrying capacity for humans on earth, the most severe global change problem will be the continued population growth in Africa and Asia (Figs. 1.1, 4.4I and 6.6). Population growth has nearly ceased in the developed world and the use of resources is beginning to decline (Figs. 6.3 and 6.6). As more people are moved above the poverty line in Africa and Asia, the population growth should decline as it has done in the developed world. However, this is a slow process and took over 50 years in America and Europe. It is unknown if this decline will be rapid enough to avoid severe problems of limiting factors and carrying capacity for the human population in Africa and Asia.

The most severe effects of global change for the humans most likely will occur in Africa. For example, Uganda has one of the highest fertility rates of 6 children per woman (Fig. 6.6) (worldpoplationreview.com). Important priorities will be to reduce poverty, increase education particularity for women, and provide techniques for family planning, which will help to slow population growth and stress on resources. Governments and private foundations in the developed world are now assisting with education and contraception in the undeveloped countries. Countries need to find solutions to cultural dogma against population control and provide more programs for population control. Otherwise millions of people will face a life of poverty, disease and premature death.

Water resources, droughts and soil loss probably will be the most important global change problem and limiting factor for the human population in the future (Figs. 4.4b, 4.14, 6.4 and 6.5) [12, 23]. Enough clean water to drink is already a problem in Africa and Asia because of the rapidly increasing populations there (Figs. 6.4 and 6.5). For example, Egypt has used 119% of renewable water resources compared to 15% in the USA (chartsbin.com). Global warming is stressing water resources everywhere, not only for humans, but also for all other of the earth's species (Figs. 4.4b, f and 6.2 Terrestrial biosphere degradation). For humans the decreasing and more polluted water resources and droughts also will reduce farmland for crops, particularly in Africa and Asia (Figs. 4.4b and 4.14) [12]. Water conservation is the most rapid method to help solve this water resources problem and this has begun in the developed world. However, there needs to be a rapid application of water resource conservation methods and reduction of polluted water in Africa and Asia.

Global warming is not only a major contributing factor to limiting water resources in the future, but it also is a major limiting and carrying capacity factor for humans and other species (Figs. 1.1, 3.11, 4.4b, f, 4.14, 6.4 and 6.5). Humans already are facing worse extreme weather events, droughts and forest fires that are more extreme and are costing more lives than in the past (e.g. Figs. 4.12 and 4.13) (e.g. Wing et al. [21]; Levi et al. 2019). How much worse the loss of life becomes in the future depends on how rapidly CO_2 and CH_4 input is limited by humans (Fig. 4.15). The Paris Agreement and similar efforts by states and cities in USA provide hope for the future, although even greater reductions in CO_2 than the presently planned reductions need to be made. The rapid introduction of solar and wind power in Europe provides an example for the greater reduction of CO_2 and CH_4. Potential reversal of the positive efforts is the rise of populist nationalism that is counter to the international and local efforts to reduce CO_2. The fate of humans lies in the hands of humans and their political leaders to have the will to limit CO_2 and CH_4 input and pollution of the air, water and soil.

Other species and entire environments face the same limiting and carrying capacity factors as humans (Fig. 1.1, 4.4f). Already there has been a significant loss of entire environments and species extinction because of human caused global warming, loss of habitat and pollution (e.g. Figs. 3.11, 4.4f and 6.2 Terrestrial biosphere degradation). Again whether entire environments such as coral reefs and mangroves survive in the future depends on human efforts to combat global change and to implement conservation efforts. Conservation and development of sustainable fisheries should remain in USA fisheries, but many fisheries in other ocean areas and lakes will disappear unless sustainable management, conservation measures and pollution control are applied.

From my aforementioned thoughts about the future, readers may think that I am too much of an alarmist. However just look at the news for one day, September 10, 2020 when worldwide scientific sources reported three significant findings for the future. The World Wildlife Fund Living Planet Report 2020 showed that the Earth's wildlife populations have fallen by an average of 68% in just the past four decades, in a decline that has not been seen for millions of years (https://livingplanet.panda.org/en-us/). The Ecological Threat Register projected that as many as 1.2 billion people around the world could be displaced by 2050 because of human-caused global changes (http://visionofhumanity.org/reports). NOAA reported that the USA hit its 10th billion-dollar weather disaster earlier than any other year. (https://www.ncdc.noaa.gov/billions/#:~:text=The%20U.S.%20has% 20sustained%20273,273%20events%20exceeds%20%241.790%20trillion). In the 1980s there were an average of 29 events per year that cost 177 billion dollars per year and from 2017 to 2020 there have been 44 events per year that have cost 460 billion dollars per year. An example of this accelerating change is that in 2019, California had 4,927 fires, which burned 118,000 acres, and by September 10th in 2020, there already had been 7,606 blazes that burned more than 3.3 million acres (fire.ca.gov).

A more positive note for the future is the rapid development of renewable energy, particularly in Europe. The USA with a populist federal government from 2017 to 2020 tried to limit development of renewable energy, however states and local governments continued their progress towards renewable energy. Eventually a greater majority of the USA population and the federal government most likely will support development of renewable energy, more conservation methods, and energy efficient transportation as the rest of the world is doing. China and India with their large populations and main reliance on burning coal have the greatest problems to solve, but they have the political will and are changing to electric vehicles more rapidly than some other countries (Figs. 6.3 and 6.6). When petroleum diminishes all countries will be forced to move rapidly toward sustainable energy.

In sum, during the foreseeable future without a meteorite hitting the earth, or nuclear holocaust, a human population probably will survive, in spite of the combined effects of all the human-caused global changes (Figs. 4.4, 6.1 and 6.2). How many and how well humans and other species survive will depend on human population growth and pollution control as well as conservation of resources and natural habitats. There will be significant reduction of natural beauty such as coral reefs, forests, and songbirds, and of recreational ski and beach areas [24, 25]. The most significant and rapid environmental changes from global warming will take place in the Arctic (Figs. 3.9, 3.10 and 4.8).

The coronavirus pandemic in 2020 is a stark reminder of how globalized the human population is and how widespread and rapid global changes can be. The pandemic shows that some countries of the world can cooperate for the common good of humanity. However, when countries or USA states with populist nationalism do not cooperate, they have had the highest number of Covid-19 cases and/or deaths (https://www.worldometers.info/coronavirus/countries-where-coronavirus-has-spread/). Part of the reason the populist countries have done so poorly with Covid-19 is their disregard for the advice of health scientists. The populist leaders should take notice of the decades old advice of the 1951 film "The Day the Earth Stood Still," in which aliens from outer space direct their warning for saving the Earth to scientists, in whom they place more hope than politicians. Humanity can be saved, aliens suggest, only if ruled by reason where an awareness of common purpose must outweigh tribal passions.

Fareed Zakaria's Global Briefing on May 14, 2020 emphasizes the importance of scientific reason and common purpose for human survival. He points out that Covid-19 has paralyzed the US, but that things could be even worse from other catastrophic threats, especially if some events combine, like natural and human-caused global changes. Some of the threats he mentions are disinformation, more pandemics, nuclear proliferation, global warming and a massive earthquake and tsunami such as on the Cascadia Subduction Zone fault (e.g. Nelson et al. 2012 [18]; Goldfinger et al. [17]; Schulz [26]). It is gratifying that our Cascadia earthquake hazard research is now on the global radar, even though little has been done to prepare for such an earthquake (e.g. https://www.opb.org/news/series/unprepared/).

The bottom line is that our future outlook can improve with a change in attitude for the common good of humanity and health of the Earth, particularly attitudes of populist politicians, the wealthy few and powerful multi-national corporations. Our choice must be to put the common good ahead of greed and short-term economic goals. This worldwide effort must be applied to the long-term human-caused global change problems of climate, pollution and diminishing natural resources, because the fate of humans is in jeopardy even more from multiple global changes than from pandemics like Covid-19.

References

1. Powell J (2019) Scientists reach 100% concensus on Anthropogenic global warming. Bull Sci Technol Soc 39:2. https://doi.org/10.1177/027046761988 6266

2. Eggers A (2019) The role of introductory geoscience courses in preparing teachers—and all students—for the future: are we making the grade? GSA Today 29:4–10

3. National Research Council (NRC) (2012) Discipline-based education research: understanding and improving learning in undergraduate science and engineering: Washington. The National Academies Press, D.C., p 282

4. NGSS Lead States (2013) Next generation science standards: for states, by states: Washington. The National Academies Press, D.C., p 324

5. Netflix (2020a) David Attenborough: A life on our planet. https://www.neflix.ctom/es-en/title/80216393

6. Gronish E (2018) Doom-and-gloom scenarios on climate change won't solve our problem, climate change: planet under pressure, Scientific American, iBooks. https://itunes.apple.com/WebObjects/MZStore.woa/wa/viewBook?id=163D2A3D07D0807D5C24F2DAB26C8E51

7. Showstack R (2018) Countries urge increased international research in the Arctic. EOS 99. https://doi.org/10.1029/2018EO108881

8. Nelson CH, Johnson KR, Barber JH (1987) Extensive whale and walrus feeding excavation on the Bering shelf, Alaska. J Sedim Petrol 57:419–430

9. Jenkins M (2019) Nature Conservancy magazine, summer 2019, 52–58

10. Jassey VEJ (2019) Effects of climate warming on Sphagnum photosynthesis in peatlands depend on peat moisture and species-specific. Glob Change Biol 25:3859–3870. https://doi.org/10.1111/gcb.14788

11. McKinsey (2019) Global Energy Perspective 2019: reference case. https://www.mckinsey.com/~/media/McKinsey/Industries/Oil%20and%20Gas/Our%20Insights/Global%20Energy%20Perspective%202019/McKinsey-Energy-Insights-Global-Energy-Perspective-2019_Reference-Case-Summary.ashx

12. Netflex (2020b) Kiss the ground. https://www.netflix.com/es-en/title/81321999

13. Holland A (2018) Preventing tomorrow's climate wars, excerpt from: Scientific American. "Climate Change: Planet Under Pressure." iBooks. https://itunes.apple.com/WebObjects/MZStore.woa/wa/viewBook?id=163D2A3D07D0807D5C24F2DAB26C8E51Scientific

14. Cartier KMS (2020) Basalts turn carbon into stone for permanent storage. EOS 101, 4. https://doi.org/10.1029/2020EO141721

15. Santer BD, Wigley TML (2007) Progress in detection and attribution research, program for climate model diagnosis and intercomparison, Lawrence Livermore National Laboratory. Livermore, CA 94550, USA. in Climate Change Science and Policy, 28 p. http://www.image.ucar.edu/idag/Papers/PapersIDAGsubtask1.4/Schneider_bookchapDnA.pdf

16. Goldfinger C, Ikeda Y, Yeats RS, Ren J (2013) Superquakes and supercycles. Seismol Res Lett 84:24–32. https://doi.org/10.1785/0220110135
17. Goldfinger C, Nelson CH, Morey AE, Gutierrez-Pastor J, Johnson JE, Karabanov E, Chaytor J, Dunhill G, Ericsson A (2012) Rupture lengths and temporal history of Cascadia great earthquakes based on turbidite stratigraphy, USGS Professional Paper 1661-F. In: Kayen R (ed) Earthquake hazards of the Pacific Northwest Coastal and Marine regions, 192 p
18. Nelson CH, Goldfinger C, Gutierrez-Pastor J (2012) Great earthquakes along the western United States continental margin: implications for hazards, stratigraphy and turbidite lithology. In: Pantosti D, Gràcia E, Lamarche G, Nelson CH (eds) Special issue on marine and lake paleoseismology. Nat Hazards Earth Syst Sci 12:3191–3208
19. Polonia A, Nelson CH, Romano S, Vaiani SC, Colizza E, Gasparotto G, Gasperini L (2017) A depositional model for seismo-turbidites in confined basins based on Ionian Sea deposits. Mar Geol 384:177–198. ISSN 0025–3227, https://doi.org/10.1016/j.margeo.2016.05.010
20. Leiserowitz A, Maibach E, Rosenthal S, Kotcher J, Bergquist P, Ballew M, Goldberg M, Gustafson A (2019) Climate change in the American mind: 2019, Yale University and George Mason University. Yale Program on Climate Change Communication. New Haven, CT, p 71. https://doi.org/10.17605/OSF.IO/CJ2NS
21. Wing OEJ, Bates PD, Smith AM, Sampson CC, Johnson KA, Fargione J, Morefield P (2018) Estimates of present and future flood risk in the conterminous United States. Environ Res Lett 13(7):034023. https://doi.org/10.1088/1748-9326/aaac65
22. Diamond J (2005) Collapse. Viking Press, p 592
23. Pimentel D (2006) Soil erosion: a food and environmental threat. J Environ Dev Sustain 8:119–137
24. Vousdoukas MI, Ranasinghe R, Mentaschi L et al (2020) Sandy coastlines under threat of erosion. Nat Clim Chang 10:260–263. https://doi.org/10.1038/s41558-020-0697-0
25. Zeng X, Boxton P, Dawson N (2018) Snowpack change from 1982 to 2016 over conterminous United States. Geophys Res Lett 45:12940–12947. https://doi.org/10.1029/2018GL079621
26. Schulz K (2015) The really big one. New Yorker. https://www.newyorker.com/magazine/2015/07/20/the-really-big-one
27. Wilson EO (2002) The future of life. Knopf Doubleday Publishing, New York City, p 220
28. United Nations (2019) Intergovernmental science-policy platform on biodiversity and ecosystem services (IPBES) Global Assessment Summary for Policymakers. www.ipbes.net/news/ipbes-global-assessment-summary-policymakers-pdf
29. United Nations (2008) World population prospects 2008
30. United Nations (2019) World population prospects, the 2008 revision

Further Reading and Information

1. American Association of Petroleum Geologists (AAPG) (2017) The Next 100 years of global energy Use: resources, impacts and economics, AAPG Search and Discovery, power point talks, Tinker #70268; Koonin #70269; Yielding #70270; Snell #70271; Ausubel #70272; Medlock III #70273

2. American Geophysical Union, EOS Earth and Space Science News. https://eos.org/

3. BBC Blue Planet II (2017) https://www.bbcearth.com/blueplanet2/#_

4. Doyle A, Fornari DJ, Brenner E, Teske AP (2019) Strategies for conducting 21st century oceanographic research. EOS 100. https://doi.org/10.1029/2019EO115729

5. Eggers A (2019) The role of introductory geoscience courses in preparing teachers—and all students—for the future: are we making the grade? GSA Today 29:4–10

6. Goldfinger C, Nelson CH, Morey AE, Gutierrez-Pastor J, Johnson JE, Karabanov E, Chaytor J, Dunhill G, Ericsson A (2012) Rupture lengths and temporal history of Cascadia great earthquakes based on turbidite stratigraphy, USGS Professional Paper 1661-F. In: Kayen R (ed) Earthquake hazards of the Pacific Northwest Coastal and Marine regions, p 184. Free download from http://pubs.usgs.gov/pp/pp1661f/

7. Heal GM (2017) Endangered economies: how the neglect of nature threatens our prosperity, Columbia University Press. New York, p 227. https://cup.columbia.edu/book/endangered-economies/9780231180849

8. IPCC (2013) Climate change 2013, The Physical Science Basis, Contribution of Working Group I to the Fifth Assessment Report of the Intergovernmental Panel on Climate Change. Stocker TF, Qin D, Plattner G-K, Tignor M, Allen SK, and others

9. IPCC (2019) Special Report on the Ocean and Cryosphere in a Changing Climate. https://www.ipcc.ch>srocc

10. Kunzig R, Broecker W (2009) Fixing climate. CPI Bookmarque Ltd., Croydon, England, p 288

11. Lansing A (1999) Endurance: Shackleton's incredible voyage, 2nd edn. Carroll & Graf Publishers. ISBN 0-7867-0621-X

12. Lee H (2020) How Earth's climate changes naturally (and why things are different now) https://www.quantamagazine.org/how-earths-climate-changes-naturally-and-why-things-are-different-now-20200721/

13. McClintock J (2012) Lost Antarctica: adventures in a disappearing land. Martins Press, St, p 256

14. McKibben W (2018) Life on a shrinking planet. The New Yorker, pp 47–55

15. McKinsey (2019) Global Energy Perspective 2019: Reference Case. https://www.mckinsey.com/~/media/McKinsey/Industries/Oil%20and%20Gas/Our%20Insights/Global%20Energy%20Perspective%202019/McKinsey-Energy-Insights-Global-Energy-Perspective-2019_Reference-Case-Summary.ashx

16. Nelson CH, Johnson KR (1987) Whales and walruses as tillers of the sea floor: Scientific American, vol 256, pp 112–117

17. Netflex (2020) Kiss the ground. https://www.netflix.com/es-en/title/81321999

18. Netflix (2019) Our Planet, Episodes 1 to 8. https://www.netflix.com/es-en/title/80049832

19. Netflix (2020) David Attenborough: a life on our planet. https://www.netflix.com/es-en/title/80216393

20. Netflix (2020) Brave Blue World https://www.netflix.com/es-en/title/81326710

21. Netflix (2021) David Attenborough Johan Rockström Breaking Boundaries: The Science Of Our Planet https://www.netflix.com/es-en/title/81336476

22. Rossi S (2019) Oceans in Decline p 352 eBook ISBN 978-3-030-02514-4 Softcover ISBN 978-3-030-02513-7

23. Schulz K (2015) The really big one. New Yorker. https://www.newyorker.com/magazine/2015/07/20/the-really-big-oneWonthePulitzerPrizeforfeaturewriting andaNationalMagazineAward

24. Scientific American (2018) Climate change: planet under pressure iBooks. https://itunes.apple.com/WebObjects/MZStore.woa/wa/viewBook?id=163D2A3D07D0807D5C24F2DAB26C8E51n

25. The Economist (2017) Special report China soil contamination, pp 5–14

26. The Economist (2019) Special report water, pp 5–14

27. The Economist (2019) Briefing the rising seas, pp 16–20

28. United Nations Intergovernmental Panel on Climate Change (IPCC). www.ipcc.ch

29. United Nations Environment Program (UNEP) Blue Carbon Initiative. https://sustainabledevelopment.un.org/partnership/?p=7405

30. United Nations Intergovernmental Science-Policy Platform on Biodiversity and Ecosystem Services (IPBES) (2019) Global assessment summary for policymakers. www.ipbes.net/news/ipbes-global-assessment-summary-policy makers-pdf

Index

Printed in the United States
by Baker & Taylor Publisher Services